Selected Titles in This Series

(Continued in the back of this publication)

The Dirichlet Problem for
Parabolic Operators with
Singular Drift Terms

MEMOIRS
of the
American Mathematical Society

Number 719

The Dirichlet Problem for Parabolic Operators with Singular Drift Terms

Steve Hofmann
John L. Lewis

May 2001 • Volume 151 • Number 719 (end of volume) • ISSN 0065-9266

American Mathematical Society
Providence, Rhode Island

2000 *Mathematics Subject Classification.*
Primary 42B20, 35K05.

Library of Congress Cataloging-in-Publication Data

Hofmann, Steve, 1958–

The Dirichlet problem for parabolic operators with singular drift terms / Steve Hofmann, John L. Lewis.

  p. cm. — (Memoirs of the American Mathematical Society, ISSN 0065-9266 ; no. 719)

"Volume 151, number 719 (end of volume)."

Includes bibliographical references.

ISBN 0-8218-2684-0

1. Dirichlet problem.   2. Parabolic operators.   I. Lewis, John L., 1943–   II. Title.   III. Series.

QA3 .A57   no. 719
[QA425]
510 s—dc21]
[515′.353]   2001018142

Memoirs of the American Mathematical Society

This journal is devoted entirely to research in pure and applied mathematics.

Subscription information. The 2001 subscription begins with volume 149 and consists of six mailings, each containing one or more numbers. Subscription prices for 2001 are $494 list, $395 institutional member. A late charge of 10% of the subscription price will be imposed on orders received from nonmembers after January 1 of the subscription year. Subscribers outside the United States and India must pay a postage surcharge of $31; subscribers in India must pay a postage surcharge of $43. Expedited delivery to destinations in North America $35; elsewhere $130. Each number may be ordered separately; *please specify number* when ordering an individual number. For prices and titles of recently released numbers, see the New Publications sections of the *Notices of the American Mathematical Society.*

Back number information. For back issues see the *AMS Catalog of Publications.*

Subscriptions and orders should be addressed to the American Mathematical Society, P. O. Box 845904, Boston, MA 02284-5904. *All orders must be accompanied by payment.* Other correspondence should be addressed to Box 6248, Providence, RI 02940-6248.

Memoirs of the American Mathematical Society is published bimonthly (each volume consisting usually of more than one number) by the American Mathematical Society at 201 Charles Street, Providence, RI 02904-2294. Periodicals postage paid at Providence, RI. Postmaster: Send address changes to Memoirs, American Mathematical Society, P. O. Box 6248, Providence, RI 02940-6248.

CONTENTS

ABSTRACT

In this memoir we consider the Dirichlet problem for parabolic operators in a half space with singular drift terms. In chapter I we begin the study of a parabolic PDE modeled on the pullback of the heat equation in certain time varying domains considered by Lewis - Murray and Hofmann - Lewis. In chapter II we obtain mutual absolute continuity of parabolic measure and Lebesgue measure on the boundary of this halfspace and also that the $L^q(R^n)$ Dirichlet problem for these PDE's has a solution when q is large enough. In chapter III we prove an analogue of a theorem of Fefferman, Kenig, and Pipher for certain parabolic PDE's with singular drift terms. Each of the chapters that comprise this memoir has its own numbering system and list of references.

1991 *Mathematics Subject Classification*. Primary 42B20, 35K05.

keywords and phrases. parabolic measure, parabolic operators, drift terms, absolute continuity, Dirichlet problem

CHAPTER I
THE DIRICHLET PROBLEM AND PARABOLIC MEASURE

1. INTRODUCTION

. The study of parabolic pde's has a long history and closely parallels the study of elliptic pde's. To mention a few highlights, the modern theory of weak solutions of elliptic and parabolic pde's in divergence form was developed in the late 1950's and early 1960's by Nash [N], DiGiorgi [DG], Moser [M], [M1], and others. These authors obtained interior estimates (boundedness, Harnack's inequality, Hölder continuity) for weak solutions which initially were assumed to lie only in a certain Sobolev space and satisfy a certain integral identity. The classical problem of whether solutions to Laplace's equation in Lipschitz domains had nontangential limits almost everywhere with respect to surface area and the corresponding L^p Dirichlet problem was not resolved until the late 70's when Dahlberg [D] showed that in a Lipschitz domain harmonic measure and surface measure, $d\sigma$, are mutually absolutely continuous, and furthermore, that the Dirichlet problem is solvable with data in $L^2(d\sigma)$. R. Hunt proposed the problem of finding an analogue of Dahlberg's result for the heat equation in domains whose boundaries are given locally as graphs of functions $\psi(x, t)$ which are Lipschitz in the space variable. It was conjectured at one time that ψ should be $\text{Lip}_{\frac{1}{2}}$ in the time variable, but subsequent counterexamples of Kaufmann and Wu [KW] showed that this condition does not suffice. Lewis and Murray [LM], made significant progress toward a solution of Hunt's question, by establishing mutual absolute continuity of caloric measure and a certain parabolic analogue of surface measure in the case that ψ has $\frac{1}{2}$ of a time derivative in $BMO(R^n)$ on rectangles, a condition only slightly stronger than $\text{Lip}_{\frac{1}{2}}$. Furthermore these authors obtained solvability of the Dirichlet problem with data in L^p, for p sufficiently large, but unspecified. Hofmann and Lewis [HL] obtained, among other results, the direct analogue of Dahlberg's theorem (i.e, L^2 solvability of the Dirichlet problem for the heat equation) in graph domains of the type considered by [LM] but only under the assumption that the above BMO norm was sufficiently small. They also provided examples to show that this smallness assumption was necessary for L^2 solvability of the Dirichlet problem.

In this memoir we study the Dirichlet problem and absolute continuity of parabolic measure for weak solutions to parabolic pde's of the form,

$$(1.1) \qquad\qquad Lu = u_t - \nabla \cdot (A\nabla u) - B\nabla u = 0.$$

Here $A = (A_{i,j}(X, t))$, $B = (B_i(X, t))$ are n by n and 1 by n matrices, respectively, satisfying standard ellipticity conditions with $X = (x_0, x_1, \ldots, x_{n-1}) = (x_0, x) \in R^n$, $t \in R$. Also ∇u denotes the gradient of u in the space variable X only, written as an n by 1 matrix, while $\nabla\cdot$ denotes divergence in the space variable. This problem in the elliptic case has been studied in [JK], [FJK], [D1] and [FKP]. As a starting point for these investigations we note that Jerison and Kenig in [JK] gave another proof of Dahlberg's results (mentioned above). To outline their proof let

---

[0]Research of both authors was supported in part by NSF grants
[0]Received by the editor June 16, 1997

$\hat{\Omega} = \{X = (x_0, x) : x_0 > \hat{\psi}(x), x \in R^{n-1}\}$, where $\hat{\psi}$ is a Lipschitz function on R^{n-1} (i.e. $|\hat{\psi}(x) - \hat{\psi}(y)| \le c|x - y|$, for some positive c, whenever $x, y \in R^{n-1}$). Let

$$\hat{\rho}(x_0, x) = (x_0 + \hat{\psi}(x), x), \ x \in R^{n-1}.$$

Then clearly $\hat{\rho}$ maps $\hat{U} = \{(x_0, x) : x_0 > 0, x \in R^{n-1}\}$ onto $\hat{\Omega}$ and $\partial\hat{U}$ onto $\partial\hat{\Omega}$ in a one to one way. If $\tilde{u}$ is a solution to Laplace's equation in $\hat{\Omega}$, then it is easily seen that $u = \tilde{u} \circ \hat{\rho}$ satisfies weakly in $\hat{U}$ a pde of the form

$$(1.2) \qquad \nabla \cdot (A\nabla u) = 0$$

where $A = A(x)$ is symmetric, satisfies standard ellipticity conditions, and has coefficients *independent of* x_0 (depending only on x). From this fact one can see that at least in spirit the pde involving A, can be differentiated with respect to x_0 to get that u_{x_0} also satisfies this pde. Using this idea and a Rellich identity, Jerison and Kenig were able to show that the Radon Nikodym derivative of harmonic measure (defined relative to (1.2) and with respect to some point in $\hat{U}$) is in a certain L^2 reverse Hölder class with respect to Lebesgue measure on $\partial\hat{U}$, whenever A is symmetric, satisfies standard ellipticity conditions and is independent of x_0.

Next we consider the analogue of this result for the heat equation in a time varying graph domain of the type considered by Lewis-Murray[LM] and Hofmann-Lewis[HL]. To this end suppose that $\psi = \psi(x, t) : R^{n-1} \times R \to R$ has compact support and satisfies

$$(1.3)$$
$$|\psi(x, t) - \psi(y, t)| \le a_1|x - y|, \text{ for some } a_1 < \infty, \text{ and all } x, y \in R^{n-1}, t \in R.$$

Also let $D^t_{1/2}\psi(x, t)$ denote the $1/2$ derivative in t of $\psi(x, \cdot), x$ fixed. This half derivative in time can be defined by way of the Fourier transform or by

$$D^t_{1/2}\psi(x, t) \equiv c \int_{\mathbb{R}} \frac{\psi(x, s) - \psi(x, t)}{|s - t|^{3/2}} \, ds$$

for properly chosen c. Assume that this half derivative exists for a.e $(x, t) \in R^n$ and $D^t_{1/2}\psi \in \mathrm{BMO}(R^n)$ with norm,

$$(1.4) \qquad \|D^t_{1/2}\psi\|_* \le a_2 < \infty.$$

Here $\mathrm{BMO}(R^n)$ (parabolic BMO) is defined as follows: Let $Q = Q_d(x, t) = \{(y, s) \in R^n : |y_i - x_i| < d, 1 \le i \le n-1, |s - t|^{1/2} < d\}$ be a rectangle in R^n and given $f : R^n \to R$, locally integrable with respect to Lebesgue n measure, let

$$f_Q = |Q|^{-1} \int_Q f(x, t) \, dxdt$$

where $dxdt$ denotes integration with respect to Lebesgue n measure and $|E|$ denotes the Lebesgue measure of the measurable set E. Then $f \in \mathrm{BMO}(R^n)$ with norm $\|f\|_*$ if and only if

$$\|f\|_* = \sup_Q \left\{|Q|^{-1} \int_Q |f - f_Q| \, dx\right\} < \infty.$$

We note that (1.3), (1.4) imply and are only slightly stronger than (1.3) and

$$(1.5)$$
$$|\psi(x, t) - \psi(x, s)| \le c(a_1 + a_2)|s - t|^{1/2} \text{ for some } c > 0 \text{ and all } x \in R^{n-1}, s, t \in R$$

(see also [HL, section 8] for the equivalence of (1.1) and (1.4) to another condition). Let $\Omega = \{(x_0, x, t) : x_0 > \psi(x, t), (x, t) \in R^n\}$ and suppose that $\tilde{u}$ is a solution to the heat equation in Ω (i.e, $\tilde{u}_t = \Delta \tilde{u}$). Let

$$\tilde{\rho}(x_0, x, t) = (x_0 + \psi(x, t), x, t), (x, t) \in R^n,$$

when $(x_0, x, t) \in U = \{(y_0, y, s) : y_0 > 0, (y, s) \in R^n\}$. Again it is easily checked that $\tilde{\rho}$ maps U onto Ω and ∂U onto $\partial \Omega$ in a 1-1 way. In this case $u = \tilde{u} \circ \tilde{\rho}$ satisfies weakly an equation of the form (1.1) where

$$B \nabla u(X, t) = \psi_t(x, t) \, u_{x_0}(X, t), \, (X, t) \in U,$$

and $A = A(x, t)$ is independent of x_0 as well as satisfies standard ellipticity conditions. Unfortunately though, $\psi_t(x, t)$ may not exist anywhere (see the remark before (1.5)). To overcome this difficulty we consider as in [HL] a transformation originally due to Dahlberg - Kenig - Stein. To this end let $\alpha \equiv (1, \ldots, 1, 2)$ be an n dimensional multi-index so that if $z = (x, t)$, then

$$\lambda^\alpha z \equiv (\lambda x, \lambda^2 t)$$

$$\lambda^{-\alpha} z \equiv (\tfrac{x}{\lambda}, \tfrac{t}{\lambda^2}).$$

Let $P(z) \in C_0^\infty(Q_1(0, 0))$ and set

$$P_\lambda(z) \equiv \lambda^{-(n+1)} P(\lambda^{-\alpha} z).$$

In addition choose $P(z)$ to be an even non-negative function, with $\int_{R^n} P(z) \, dz \equiv 1$. Next let $P_\lambda \psi$ be the convolution operator

$$P_\lambda \psi(z) \equiv \int_{R^n} P_\lambda(z - v) \psi(v) \, dv.$$

and put

(1.6) $\rho(x_0, x, t) = (x_0 + P_{\gamma x_0} \psi(x, t), x, t), \text{ when } (x_0, x, t) \in U.$

From properties of parabolic approximate identities and (1.3), (1.5), it is easily checked that $\lim\limits_{(y_0, y, s) \to (x, t)} P_{\gamma y_0} \psi(y, s) = \psi(x, t)$. Thus ρ extends continuously to ∂U. Also if γ is small enough (depending on a_1, a_2), it is easily shown that ρ maps U onto Ω and ∂U onto $\partial \Omega$ in a one to one way. Next observe that if $\tilde{u}$ is a solution to the heat equation in Ω, then $u = \tilde{u} \circ \rho$ is a weak solution to an equation of the form (1.1) where A satisfies standard ellipticity estimates. Before proceeding further we note that parabolic measure on ∂U, defined with respect to this pullback pde and a point in U, is absolutely continuous with respect to Lebesgue measure on ∂U, thanks to [LM, ch.3] (in fact parabolic measure defined with respect to a given point is an A_∞ weight with respect to Lebesgue measure on a certain rectangle). Thus the pullback pde should be a good model for proving mutual absolute continuity of parabolic and Lebesgue measure.

In chapters I and II of this memoir we study the remarkable structure of this pullback pde. In chapter I we establish certain basic estimates for parabolic pde's with singular drift terms and establish L^2 solvability of the Dirichlet problem for pde's which are near a constant coefficient pde in a certain Carleson measure sense. In chapter II we remove the nearness assumption on the Carleson measures considered in chapter I and thus obtain our first main theorem on absolute continuity of parabolic measure and the corresponding L^q Dirichlet problem. As a corollary we obtain the results of [LM] mentioned above. In chapter III we obtain parabolic analogues for pde's with singular drift terms of theorems in [FKP].

We emphasize that our results are not straight forward generalizations of theorems for elliptic equations. For example we do not know if the pde's we consider in chapter 2 have parabolic measures which are doubling, as is well known for the corresponding elliptic measures. Also we cannot prove certain basic estimates such as Hölder continuity for the adjoint Green's function of our pde's. In this respect our work is more akin to results of [VV] and [KKPT] in two dimensions. Finally we mention that the possible lack of doubling for our parabolic measures forces us in chapter III to give alternative arguments in place of the usual square function - nontangential maximum arguments.

The first author would like to thank Carlos Kenig for helpful discussions concerning necessary conditions on A, B to prove Theorem 2.13. The second author would like to thank Russell Brown and Wei Hu for useful discussions concerning basic estimates for pde's with drift terms.

2. STATEMENT OF RESULTS

As rationale for the structure assumptions on our pde's, we shall briefly outline the structure of the pullback pde under the mapping given in (1.6). To this end recall that a positive measure μ is said to be a Carleson measure on U if for some positive $c < \infty$

$$\mu[(0, d) \times Q_d(x, t)] \leq c|Q_d(x, t)| \text{ for all } d > 0, \ (x, t) \in R^n.$$

The infimum over all c for which the above inequality holds is called the Carleson norm of μ and denoted $\|\mu\|$. The following lemma is proved in [HL, Lemma 2.8].

Lemma A. *Let σ, θ be nonnegative integers and $\phi = (\phi_1, \ldots, \phi_{n-1})$, a multi-index, with $l = \sigma + |\phi| + \theta$. If ψ satisfies (1.3), (1.4) for some $a_1, a_2 < \infty$, then the measure ν defined at (x_0, x, t) by*

$$d\nu = \left(\frac{\partial^l P_{\gamma x_0} \psi}{\partial x_0^\sigma \partial x^\phi \partial t^\theta} \right)^2 x_0^{(2l+2\theta-3)} \, dx dt dx_0$$

is a Carleson measure whenever either $\sigma + \theta \geq 1$ or $|\phi| \geq 2$, with

$$\nu[(0, d) \times Q_d(x, t)] \leq c \, |Q_d(x, t)| \, .$$

Moreover, if $l \geq 1$, then at (x_0, x, t)

$$| \frac{\partial^l P_{\gamma x_0} \psi}{\partial x_0^\sigma \partial x^\phi \partial t^\theta} | \leq c' \, (a_1 + a_2) \, x_0^{1-l-\theta}$$

where $c' = c'(n)$ and $c - c(a_1, a_2, \gamma, l, n) \geq 1$.

Remark. The last inequality in Lemma A remains true under the weaker assumptions (1.3), (1.5). We shall use this remark in chapters II and III.

Recall that in section 1 we introduced the pullback function, $u = \tilde{u} \circ \rho$, where ρ is as in (1.6). Also u satisfied a certain pullback pde of the form (1.1). We note that a typical term in the pullback drag term $B \, \nabla u$, evaluated at (X, t), is $\frac{\partial}{\partial t} P_{\gamma x_0} \psi \, u_{x_0}$. From Lemma A with $\sigma = 0 = |\phi|$, $\theta = 1$, we see that

$$d\mu(X, t) = x_0 [\frac{\partial}{\partial t} P_{\gamma x_0} \psi(x, t)]^2 \, dX dt$$

is a Carleson measure on U. Thus a natural assumption on B is that

$$d\mu_1(X,t) = x_0 |B|^2(X,t) \, dXdt,$$

is a Carleson measure on U with

(2.1) $$\|\mu_1\| \leq \beta_1 < \infty.$$

Next observe from the above lemma with $\theta = |\phi| = 0$, $\sigma = 1$, that

$$x_0^{-1}[\tfrac{\partial}{\partial x_0} P_{\gamma x_0} \psi(x,t)]^2 \, dXdt$$

is a Carleson measure on U. Unfortunately a typical term in $A\nabla u$ evaluated at (X,t) is $[\tfrac{\partial}{\partial x_i} P_{\gamma x_0} \psi]^2 \, u_{x_0}$, $1 \leq i \leq n-1$, and for each such i, the measure with density

$$x_0^{-1}[\tfrac{\partial}{\partial x_i} P_{\gamma x_0} \psi]^2 \, dXdt$$

need not give rise to a Carleson measure. The failure of this measure to be Carleson makes the structure of A for the pullback pde complicated and causes us to make an abundance of assumptions on A (all are needed in the estimates and all are satisfied by our model term, as can be deduced from Lemma A). First assume that

(2.2) $$(x_0 |\nabla A| + x_0^2 |A_t|)(X,t) < \Lambda < \infty$$

for a.e $(X,t) \in U$ and if

$$d\mu_2(X,t) = (x_0 |\nabla A|^2 + x_0^3 |A_t|^2)(X,t) \, dXdt,$$

then μ_2 is a Carleson measure on U with

(2.3) $$\|\mu_2\| \leq \beta_2 < \infty.$$

Second assume that whenever $0 \leq i,j \leq n-1$, we have

$$\frac{\partial A_{ij}}{\partial x_0} = \sum_{l=0}^{n-1} \langle e_l^{ij}, \tfrac{\partial}{\partial x_l} f_l^{ij} \rangle + g^{ij}$$

in the distributional sense. Here

$$e_l^{ij} = (e_{l1}^{ij}, e_{l2}^{ij}, \dots, e_{ln_l}^{ij}),$$

$$f_l^{ij} = (f_{l1}^{ij}, f_{l2}^{ij}, \dots, f_{ln_l}^{ij}),$$

are measurable functions from $U \to R^{n_l}$ for $0 \leq l \leq n-1$ and $\langle e_l^{ij}, f_l^{ij} \rangle$ denotes the inner product of these functions as vectors in R^{n_l}. Third assume that

(2.4) $$[\sum_{l=0}^{n-1} |e_l^{ij}| + |f_l^{ij}|](X,t) \leq \Lambda < \infty$$

and that e_l^{ij} has distributional first partials in X whenever $0 \leq i,j \leq n-1$. In (2.4), $|e_l^{ij}|$, $|f_l^{ij}|$ denote the norm of these functions considered as vectors in R^{n_l}. Let ∇e_l^{ij} denote the gradient of e_l^{ij} taken componentwise. Fourth assume that

$$d\mu_3(X,t) = [\sum_{i,j=0}^{n-1} (\sum_{l=0}^{n-1} x_0 |\nabla e_l^{ij}|^2 + x_0^{-1} |f_l^{ij}|^2) + |g^{ij}|](X,t) \, dXdt$$

is a Carleson measure on U with

(2.5) $$\|\mu_3\| \leq \beta_3 < \infty.$$

Under these conditions and standard ellipticity assumptions, we shall show the Radon-Nikodym derivative of parabolic measure on rectangles is in a certain reverse Hölder class when $\beta_i, 1 \leq i \leq 3$, are small.

In order to state the main theorem in chapter I precisely we introduce some notation which will be used throughout this memoir. For completeness we restate some of the notation used earlier. Let $G, \partial G, |G|$, denote the closure, boundary, and Lebesgue $n, n + 1$ measure of the set G, whenever $G \subset R^n$ or R^{n+1} and there is no chance of confusion. If $G \subset R^n$, let $L^p(G), 1 \leq p \leq \infty$, be the space of equivalence classes of Lebesgue measurable functions f on G which are p th power integrable with norm denoted by $\|f\|_{L^p(G)}$. If G is open let $C_0^\infty(G)$ be infinitely differentiable functions with compact support in G. For k a positive integer let $H^k(G)$ be the Sobolev space of equivalence classes f whose distributional partial derivatives $D^\beta f$ ($\beta = (\beta_0, \beta_1, \ldots, \beta_{n-1}) = $ multi - index) of order $\leq k$ are square integrable. Let

$$\|f\|_{H^k(G)} = \left\| \left(\sum_{|\beta| \leq k} |D^\beta f|^2 \right)^{1/2} \right\|_{L^2(G)}$$

and put $H_0^k(U)$ equal to the closure in $C_0^\infty(U)$ of $H^k(U)$. We say that $f \in H_{\mathrm{loc}}^k(G)$, $L_{\mathrm{loc}}^p(G)$, if $f \in H^k(G_1), L^p(G_1)$, respectively, whenever G_1 is open with $\bar{G}_1 \subset G$. Let $L^p(T_1, T_2, H_{\mathrm{loc}}^k(G)), 1 \leq p \leq \infty, k$ a positive integer, be equivalence classes of Lebesgue measurable functions $f : G \times (T_1, T_2) \to R$ with $f(\cdot, t) \in H_{\mathrm{loc}}^k(G)$ for almost every $t \in (T_1, T_2)$ and

$$\int_{T_1}^{T_2} \|f(\cdot, t)\|_{H^k(G_1)}^p \, dt < \infty,$$

whenever G_1 is open with $\bar{G}_1 \subset G$. $L^p(T_1, T_2, L_{\mathrm{loc}}^p(G))$ is defined similarly with $H_{\mathrm{loc}}^k(G)$ replaced by $L_{\mathrm{loc}}^p(G)$.

As introduced earlier, $\nabla = (\frac{\partial}{\partial x_0}, \ldots, \frac{\partial}{\partial x_{n-1}})$ while $\nabla \cdot = \sum_{i=0}^{n-1} \frac{\partial}{\partial x_i}$. Unless otherwise stated c will denote a positive constant depending only on the dimension, not necessarily the same at each occurence, while $c(\beta, \mu, \nu)$ will denote a constant depending only on β, μ, ν. Also points in R^{n+1} will be denoted by (X, t) or (x_0, x, t) while $Q_d(x, t) \subset R^n$ will denote the rectangle with center (x, t), side length $2d$ in the space variables, and side length $2d^2$ in the time variable. We write

$$Q_d(X, t) = (x_0 - d, x_0 + d) \times Q_d(x, t) \subset R^{n+1}$$

when there is no chance of confusion. Let $\beta_p(Q_d(x, t))$ be the reverse Hölder class of functions $f : R^n \to R$ with $\|f\|_{L^p(Q_d(x,t))} < \infty$ and

$$(2.6) \qquad |Q_r(y, s)|^{-1} \int_{Q_r(y,s)} f^p \, dx dt \leq \lambda^p \left(|Q_r(y, s)|^{-1} \int_{Q_r(y,s)} f \, dx dt \right)^p$$

for some $\lambda, 0 < \lambda < \infty$, and all rectangles with $Q_r(y, s) \subset Q_d(x, t)$. Let $\|f\|_{\beta_p(Q_d(x,t))}$ be the infimum of the set of all λ such that (2.6) holds. Similarly, let $\alpha_p(Q_d(x, t))$ be the underline{weak} reverse Hölder class of functions f defined as above except

$$(2.7) \qquad |Q_r(y, s)|^{-1} \int_{Q_r(y,s)} f^p \, dx dt \leq \lambda^p \left(|Q_{2r}(y, s)|^{-1} \int_{Q_{2r}(y,s)} f \, dx dt \right)^p$$

for some $\lambda, 0 < \lambda < \infty$, and all rectangles with $Q_{2r}(y,s) \subset Q_d(x,t)$. Let $\|f\|_{\alpha_p(Q_d(x,t))}$ be the infimum of the set of all λ such that (2.7) holds.

Next let $A = (A_{ij}(X,t)), 0 \leq i,j \leq n-1$, $B = (B_i(X,t)), 0 \leq i \leq n-1$, be the $n \times n$ and $1 \times n$ matrices defined in section 1. We assume that $A_{ij}, B_i : U \to R$ are Lebesgue measurable and that A satisfies the standard ellipticity condition,

$$(2.8) \qquad \langle A(X,t)\xi, \xi \rangle \geq \gamma_1 |\xi|^2$$

for some $\gamma_1 > 0$, almost every $(X,t) \in U$ and all $n \times 1$ matrices ξ. Here $\langle \cdot, \cdot \rangle$ denotes the usual inner product on R^n. We also assume that

$$(2.9) \qquad \left(\sum_{i=0}^{n-1} x_0 |B_i| + \sum_{i,j=0}^{n-1} |A_{ij}| \right)(X,t) < M < \infty$$

for almost every $(X,t) \in U$. To simplify matters we shall assume for some large $\rho > 0$, that

$$(2.10) \qquad A \equiv \text{ constant matrix in } U \setminus Q_\rho(0,0).$$

Following Aronsson [A] we say that u is a weak solution to (1.1) in U if for $\hat{U} = \{(x_0, x) : x \in R^{n-1}, x_0 > 0\}$, $-\infty < T < \infty$, we have

$$(2.11) \qquad u \in L^2(-T, T, H^1_{\text{loc}}(\hat{U})) \cap L^\infty(-T, T, L^2_{\text{loc}}(\hat{U}))$$

and

$$(2.12) \qquad \int_U \left(-u\phi_t + \sum_{i,j=0}^{n-1} A_{ij} u_{x_j} \phi_{x_i} - \sum_{i=0}^{n-1} B_i u_{x_i} \phi \right) dXdt = 0$$

for all $\phi \in C_0^\infty(U)$.

In the sequel we shall identify ∂U with R^n. The continuous Dirichlet problem for U can be stated as follows: Given $g : R^n \to R$, continuous, and bounded, find u a bounded weak solution to (1.1) in U with u continuous on $\bar{U}$ and $u = g$ on ∂U. Assume that the Dirichlet problem for a given A, B always has a unique solution. Under this asumption we define parabolic measure ω at $(X,t) \in U$ of the Borel measurable set $E \subset R^n$ by

$$\omega(X, t, E) = \inf \{v(X,t) : v \in \mathcal{F}\}$$

where $\mathcal{F}$ denotes the family of all nonnegative solutions to the Dirichlet problem in U with $v \geq 1$ on E. Finally let $\frac{d\omega}{dyds}$ denote the Radon-Nikodym derivative of ω with respect to Lebesgue measure on R^n. With this notation we are now ready to state the main theorem in chapter I.

Theorem 2.13. *Let A, B, satisfy (2.1)-(2.5) and (2.8)-(2.11). Suppose for some $\epsilon_0 > 0$ and A_0 an $n \times n$ matrix that*

$$\||x_0 B|\|^2_{L^\infty(U)} + \||A - A_0|\|^2_{L^\infty(U)} + \|\mu_1\| + \|\mu_2\| + \|\mu_3\| \leq \epsilon_0^2.$$

If $\epsilon_0 > 0$ is small enough, then the continuous Dirichlet problem corresponding to (1.1), A, B, always has a unique solution. If ω denotes the corresponding parabolic measure, then $\omega(d, x, t + 2d^2, \cdot)$ is mutually absolutely continuous with respect to Lebesgue measure on $Q_d(x,t)$ and $\frac{d\omega}{dyds}(d, x, t + 2d^2, \cdot) \in \beta_2(Q_d(x,t))$ with

$$\|\tfrac{d\omega}{dyds}(d, x, t + 2d^2, \cdot)\|_{\beta_2(Q_d(x,t))} < c^* < \infty,$$

for all $(x, t) \in \mathbf{R}^n$, $d > 0$. *Here* $c^* = c^*(\epsilon_0, \gamma_1, M, \Lambda, n)$.

Remark. 1) In Theorem 2.13, $x_0 B$ denotes the 1 by n matrix function, $(X, t) \to x_0 B(X, t)$. We note that the smallness assumption in Theorem 2.13 can be weakened. We do not prove this weakened version since its proof is more complicated and since we are primarily interested in the case when $\|\mu_1\| + \|\mu_2\| + \|\mu_3\|$ is large. We refer the reader to the remark at the end of section 5 for an exact statement of a stronger form of Theorem 2.13.

2) To prove the above result for small $\epsilon_0 > 0$, we shall use local estimates in [A], [M], [M1], and [FGS] for solutions to the pde in (1.1) and its adjoint pde, but we will also need to show that if a solution to (1.1) or its adjoint pde has continuous zero boundary value on $Q_{2d}(x, t) \subset \partial U$, then this solution is Hölder continuous on $(0, d) \times Q_d(x, t)$. The proof of Hölder continuity cannot be deduced from the usual arguments (such as reflection) since the drag term B evaluated at (X, t) can blow up as $x_0 \to 0$ (almost like x_0^{-1}) even under the above Carleson measure assumptions on B. Using these basic estimates it is not difficult to show that the continuous Dirichlet problem for the pde in (1.1) always has a unique solution. Moreover, we can use these estimates to modify slightly an argument of Fabes and Safonov [FS] to show first that the adjoint Green's function corresponding to (1.1) satisfies a backward Harnack inequality in U when ϵ_0 is sufficiently small and second that parabolic measure corresponding to (1.1) is a doubling measure. To be more precise, we show that

$$\omega(d, x, t + 2d^2, Q_{2r}(y, s)) \le c\, \omega(d, x, t + 2d^2, Q_r(y, s))$$

whenever $(x, t) \in \mathbf{R}^n, d > 0$, and $Q_{2r}(y, s) \subset Q_d(x, t)$ provided ϵ_0 is sufficiently small. We shall make our basic estimates and prove doubling for pde's of the form (1.1) in section 3.

In section 4 we begin the proof of Theorem 2.13. We first show that parabolic measure is in the above reverse Hölder class when $B \equiv 0$ and all the above Carleson norms are small. In this case we perturb our results off a constant coefficient pde by making estimates of the form:

$$\left| \int\!\!\!\int_{Q_{x_0}(X, t)} (A - A_0)_{00}\, G_{y_0} h_{y_0}\, dY\, ds \right| \le \delta\, \|N(|\nabla G|)\|_{L^2(Q_{2x_0}(x, t))}\, \|S(h)\|_{L^2(\mathbf{R}^n)},$$

where $\delta \to 0$ as $\epsilon_0 \to 0$ while G is the Green's function for (1.1) with $B \equiv 0$ and pole at $(x_0, x, t + 2x_0^2)$. Also h is a weak solution to $\nabla \cdot (A_0 \nabla h) = 0$ while N, S are defined below in (2.14), (2.15). We note that if

$$y_0^{-1} |(A - A_0)_{00}|^2 (Y, s)\, dY\, ds$$

were a Carleson measure on U, then the above estimate would be an easy consequence of Cauchy's inequality and (2.16) at the end of this section. Since this measure need not be Carleson (compare with the prototype equation discussed in section 1), we are forced to integrate by parts numerous times in x_0, x, t and use all of our Carleson measure assumptions on A, in order to obtain the above estimate. The case when $B \not\equiv 0$ and ϵ_0 is small, follows easily from the above case using our basic estimates and another perturbation type argument. The proof of Theorem 2.13 is given in sections 4 and 5.

We close this section by defining the nontangential maximal and square functions introduced above. Given $a > 0$ and $(x,t) \in R^n$, let

$$\Gamma(x,t) = \Gamma_a(x,t) = \{(Y,s) \in U : (y,s) \in Q_{ay_0}(x,t)\}$$

and if $g : U \rightarrow R$, put

(2.14)
$$Ng(x,t) = \sup_{(Y,s) \in \Gamma_a(x,t)} |g|(Y,s).$$

If g has a locally integrable distributional gradient, ∇g, on U let

(2.15)
$$Sg(x,t) = \left(\int_{\Gamma_a(x,t)} y_0^{-n} |\nabla g|^2 (Y,s) \, dY \, ds \right)^{1/2}.$$

Ng and Sg are called the nontangential maximal function and area function of g defined relative to $\Gamma_a(x,t)$. Finally for g as above and μ a Carleson measure on U we note (see [St, p 236]) that for $1 \le p < \infty$

(2.16)
$$\int_U |g|^p \, d\mu \le c(p,a,n) \|\mu\| \, \|Ng\|^p_{L^p(\mathbb{R}^n)}.$$

3. BASIC ESTIMATES

In this section we state some basic estimates from [M1, M2], [A], and [FGS] for weak solutions to (1.1) when A, B satisfy (2.8)-(2.10). We shall also need basic estimates for weak solutions v to the adjoint pde in U corresponding to (1.1), i.e.

(3.1)
$$\tilde{L}v = v_t + \nabla \cdot (A^\tau \nabla v - Bv) = 0,$$

where A^τ is the transpose matrix corresponding to A. More specifically, we have

$$v \in L^2(-T,T,H^1_{\text{loc}}(\hat{U})) \cap L^\infty(-T,T,L^2_{\text{loc}}(\hat{U}))$$

for $0 < T < \infty$ and

(3.2)
$$\int_U \left(v\phi_t + \sum_{i,j=0}^{n-1} A_{ij} v_{x_j} \phi_{x_i} - v \sum_{i=0}^{n-1} B_i \phi_{x_i} \right) dX \, dt = 0$$

for all $\phi \in C_0^\infty(U)$. Weak solutions to (1.1) or (3.1) in $Q_d(X,t)$ are defined similarly to weak solutions in U except that $\hat{U}$ is replaced by $\{Y : |x_i - y_i| < d, \, 0 \le i \le n-1\}$. We say that v is a local solution to (1.1) or (3.1) in an open set O, if v is a weak solution in each $Q_d(X,t) \subset O$. We shall need the following interior estimates (see [A, section 2]).

Lemma 3.3 (Parabolic Cacciopoli). *Let A, B, satisfy (2.8)-(2.10) and suppose that u is a weak solution to either (1.1) or (3.1) in $Q_{4d}(X,t)$, $0 < d < x_0/8$. If $Q(s) = Q_d(X,t) \cap (R^n \times \{s\})$, then*

$$d^n \left(\max_{Q_{d/2}(X,t)} u \right)^2 \le c \sup_{s \in (t-d^2,\, t+d^2)} \int_{Q(s)} u^2(Y,s) \, dY$$

$$+ c \int_{Q_d(X,t)} |\nabla u|^2 (Y,s) \, dY \, ds$$

$$\le c^2 \, d^{-2} \int_{Q_{2d}(X,t)} u^2(Y,s) \, dY \, ds$$

for some $c = c(\gamma_1, M, n) \geq 1$, where γ_1, M, are as in (2.8)-(2.9).

Lemma 3.4 (Interior Hölder Continuity). *Let A, B, satisfy (2.8)-(2.10) and suppose that u is a weak solution to either (1.1) or (3.1) in $Q_{4d}(X, t)$, $0 < d < x_0/8$. If $|u| \leq K < \infty$ in $Q_{4d}(X, t)$ and $(Y, s), (Z, \tau) \in Q_{2d}(X, t)$, then*

$$|u(Y, s) - u(Z, \tau)| \leq c\, K \left(\frac{|Y - Z| + |s - \tau|^{1/2}}{d} \right)^\alpha$$

for some $c = c(\gamma_1, M, n)$, $\alpha = \alpha(\gamma_1, M, n)$, $0 < \alpha < 1 \leq c < \infty$.

Lemma 3.5 (Harnack's Inequality). *Let A, B, satisfy (2.8)-(2.10) and suppose that $(Y, s), (Z, \tau) \in Q_{2d}(X, t)$. There exists $c = c(\gamma_1, M, n)$ such that if $u \geq 0$ is a weak solution to (1.1) in $Q_{4d}(X, t)$, $0 < d < x_0/8$, then for $\tau < s$,*

$$u(Z, \tau) \leq u(Y, s) \exp\left[c \left(\frac{|Y - Z|^2}{|s - \tau|} + 1 \right) \right]$$

while if $u \geq 0$ is a weak solution to (3.1), then this inequality is valid when $\tau > s$.

Remark. In [A] the above lemmas are stated for $\||B\||_{L^\infty(U)} < c < \infty$. However using the scaling $(X, t) \to (X/d, t/d^2)$, we can reduce the proof of Lemmas 3.3 - 3.5 to a rectangle of side length 1 in the space variable that is distance 1 from ∂U. From (2.9) we see that $|B|$ is bounded almost everywhere on such a rectangle, so the results in [A] can be used.

We suppose until further notice that either $B \equiv 0$ in (1.1) or $A, B \in C^\infty(\bar{U})$. Under these assumptions there exists Green's function G for (1.1) in U and corresponding parabolic, adjoint parabolic measures $\omega(X, t, \cdot), \tilde{\omega}(X, t, \cdot)$, satisfying for each $(X, t) \in U$ the condition:

$$\phi(X, t) = \int_U \left[\langle A\nabla\phi, \nabla_Y\, G(X, t, \cdot) \rangle + G(X, t, \cdot)\, (\phi_s - B\,\nabla\phi) \right] dY\, ds$$

(3.6a)

$$+ \int_{\partial U} \phi(y, s)\, d\omega(X, t, y, s),$$

(3.6b)
$$\phi(X, t) = \int_U \left\{ \langle A^\tau\nabla\phi, \nabla_Y\, G(\cdot, X, t) \rangle + G(\cdot, X, t)\, (-\phi_s + \nabla \cdot [B\phi]) \right\} dY\, ds$$

$$+ \int_{\partial U} \phi(y, s)\, d\tilde{\omega}(X, t, y, s),$$

whenever $\phi \in C_0^\infty(R^{n+1})$. Moreover, G has the properties:

(3.7)

(a) $G(X, t, Y, s) = 0$ for $s > t$, (X, t), $(Y, s) \in U$,

(b)   For fixed $(Y, s) \in U$, $G(\cdot, Y, s)$ is a local solution to (1.1) in $U \setminus \{(Y, s)\}$,

(c)   For fixed $(X, t) \in U$, $G(X, t, \cdot)$ is a local solution to (3.1) in $U \setminus \{(X, t))\}$,

(d)   If (X, t), $(Y, s) \in U$, then $G(X, t, \cdot)$ and $G(\cdot, Y, s)$ extend continuously to $\bar{U}$ provided both functions are defined to be zero on ∂U.

We note that (3.6), (3.7) are well known when A, B are smooth (see [F]) while if $B \equiv 0$ a proof can be given by taking weak limits of solutions to pde's with smooth coefficients. It is also easily seen (as in [A]) that the solution to the Dirichlet problem for (1.1) in U and a given bounded continuous $h : R^n \to R$ is given by

$$(3.8) \qquad u(X, t) = \int_{\mathbb{R}^n} h(y, s) \, d\omega(X, t, y, s).$$

Next we state two lemmas from [FGS] concerning the Green's function and parabolic measure. In [FGS] these lemmas are given in Lipschitz cylinders. However these lemmas remain valid for U as is easily seen.

Lemma 3.9 (Boundary Hölder Continuity).   *Let A satisfy (2.8)-(2.10), $B \equiv 0$, and let u be a weak solution to (1.1) or (3.1) in $(0, 2r) \times Q_{2r}(y, s)$. If $r > 0$ and u vanishes continuously on $Q_{2r}(y, s)$, then there exists $c = c(\gamma_1, M, n)$ and $\alpha = \alpha(\gamma_1, M, n)$, $0 < \alpha < 1 \leq c < \infty$, such that*

$$u(X, t) \leq c \, (x_0/r)^\alpha \max_{(0, r) \times Q_r(y, s)} u$$

whenever $(X, t) \in (0, r/2) \times Q_{r/2}(y, s)$. If $u \geq 0$ in $(0, 2r) \times Q_{2r}(y, s)$, then there exists $\tilde{c} = \tilde{c}(\gamma_1, M, n)$ such that for (X, t) as above,

$$u(X, t) \leq \tilde{c} \, (x_0/r)^\alpha \, u(r, y, s \pm 2r^2)$$

where the plus sign is taken when u is a weak solution to (1.1) and the minus sign for u satisfying (3.1).

Lemma 3.10.   *Let G, ω be Green's function and parabolic measure corresponding to (1.1) in U where A satisfies (2.8)-(2.10) and $B \equiv 0$. If $(y, s) \in R^n$, $r > 0$, then there exists $c \geq 1$, depending on γ_1, M, n, such that*

$$c^{-1} \, r^n \, G(X, t, r, y, s + 100r^2) \leq \omega(X, t, Q_r(y, s))$$

$$\leq c \, \omega(X, t, Q_{2r}(y, s)) \leq c^2 \, r^n \, G(X, t, r, y, s - 100r^2),$$

whenever, $t \geq s + 200r^2$, $x_0 \geq 4r$, and

$$10^{-3n} \leq \frac{|X - (r, y)|^2}{t - s} + \frac{t - s}{x_0^2} \leq 10^{3n}.$$

We shall also need the following backward Harnack inequality in [FS].

Lemma 3.11 (Backward Harnack Inequality). *Let G be Green's function for (1.1) in U with A satisfying (2.8)-(2.10) and $B \equiv 0$. There exists $c = c(\gamma_1, M, n) \geq 1$ such that for $(X, t), (y, s), r$, as in Lemma 3.10,*

$$G(X, t, r, y, s - 100r^2) \leq c\, G(X, t, r, y, s + 100r^2).$$

This backward Harnack inequality is proven in [FS] for weak solutions to (1.1) in bounded Lipschitz cylinders. However using Lemmas 3.5, 3.9, 3.10, it is easily seen that their proof extends to the situation in Lemma 3.11. From Lemmas 3.5, 3.10, 3.11, we conclude

Lemma 3.12 (Parabolic Doubling). *Let ω be parabolic measure corresponding to (1.1) in U with A satisfying (2.8)-(2.10) and $B \equiv 0$. There exists $c = c(\gamma_1, M, n) \geq 1$ such that*

$$\omega(X, t, Q_{2r}(y, s)) \leq c\, \omega(X, t, Q_r(y, s))$$

where $(X, t), (y, s), r$ are as in Lemma 3.10.

Next we show that Lemmas 3.9 - 3.12 remain valid when $B \not\equiv 0$ provided for almost every $(X, t) \in U$, we have

(3.13)
$$x_0 \sum_{i=0}^{n-1} |B_i(X, t)| \leq \epsilon_1$$

and ϵ_1 is sufficiently small. We prove

Lemma 3.14. *Let $A, B \in C^\infty(\bar{U})$ satisfy (2.8)-(2.10) and (3.13). If $\epsilon_1 = \epsilon_1(\gamma_1, M, n), 0 < \epsilon_1 < 1$, is small enough, then Lemmas 3.9 - 3.12 are still true.*

Remark. If $u(X, t) = [-\log(x_0)]^{-1}$, then $\Delta u + B\nabla u = 0$ for $0 < x_0 < 1/2$ where $B = (x_0^{-1}[1 + \frac{2}{\log x_0}], 0, \ldots, 0)$. Thus some smallness condition such as (3.13) is needed in order to insure the validity of Lemma 3.9.

Proof of Lemma 3.9 : We first extend u to R^{n+1} by defining $u \equiv 0$ in $R^{n+1} \setminus [(0, 2r) \times Q_{2r}(y, s)]$. Let $0 < \rho < r$, $(y_1, s_1) \in Q_{r/2}(y, s)$, and $\psi \geq 0 \in C_0^\infty((-\rho, \rho) \times Q_\rho(y_1, s_1))$ with $\psi \equiv 1$ on $(-\rho/2, \rho/2) \times Q_{\rho/2}(y_1, s_1)$. Also choose ψ so that

$$\rho \, \| \, |\nabla \psi| \, \|_{L^\infty(\mathbb{R}^{n+1})} + \rho^2 \, \|\tfrac{\partial}{\partial t}\psi\|_{L^\infty(\mathbb{R}^{n+1})} \leq c.$$

If u vanishes continuously on $Q_{2r}(y, s)$, then from Schauder type estimates (see[F]), we deduce that $\phi = u\psi^2$ can be used as a test function in (2.12) or (3.2). Using this test function we obtain for some $c = c(\gamma_1, M, n) \geq 1$ that

$$I_1 = c^{-1} \int_U |\nabla u|^2 \, \psi^2 \, dX dt \leq \int_U \langle \tilde{A} \nabla u, \nabla u \rangle \, \psi^2 \, dX dt$$

$$\leq c \int_U |u| \, |\nabla u| \, \psi \, |\nabla \psi| \, dX dt + c |\int_U u \, \tfrac{\partial}{\partial t}(u \psi^2) dX dt|$$

(3.15)

$$+ c \int_U (\, |u| \, |B| \, |\nabla u| \, \psi^2 + u^2 \, |B| \, |\nabla \psi| \, \psi \,) dX dt$$

$$= I_2 + I_3 + I_4 \, .$$

Here $\tilde{A} = A$ when u is a weak solution to (1.1) while $\tilde{A} = A^\tau$ when u is a weak solution to (3.1). Also $|B|$ denotes the norm of B considered as a vector. Using Cauchy's inequality with ϵ's we find in the usual way that

$$I_2 \leq \tfrac{1}{2} I_1 + c \int_U u^2 \, |\nabla \psi|^2 \, dX dt.$$

Differentiating the product in I_3 and then integrating the term in $\frac{\partial}{\partial t} u^2$ by parts one gets

$$I_3 \leq c \int_U u^2 \, | \tfrac{\partial}{\partial t} \psi^2 \, | \, dX dt$$

To estimate I_4 we use the fact that for $(X, t) \in (0, r) \times Q_r(y, s)$ we have

(3.16) $$u \psi(X, t) \leq x_0 \, M^{(1)}[|\nabla(u\psi)|](X, t)$$

where

$$M^{(1)} f(X, t) = x_0^{-1} \int_0^{x_0} |f|(z_0, x, t) \, dz_0$$

is $\leq$ the one dimensional maximal function in the x_0 variable when (x, t) are held constant. Using (3.16), (3.13), Cauchy's inequality, and the Hardy - Littlewood maximal theorem we get

$$I_4 \leq c \epsilon_1 \int_U u^2 \, |\nabla \psi|^2 \, dX dt + c \epsilon_1 \int_U |\nabla u|^2 \, \psi^2 \, dX dt.$$

Putting these estimates for I_2, I_3, I_4 in (3.15) we see for ϵ_1 small enough that

(3.17) $$I_1 \leq c \int_U \left(|\nabla \psi|^2 + |\tfrac{\partial}{\partial t} \psi| \right) u^2 \, dX dt.$$

From (3.17) and (3.16) with $u\psi$ replaced by u, as well as our assumptions on ψ, we conclude for $\epsilon_1 = \epsilon_1(\gamma_1, M, n)$ small enough that there exists $c = c(\gamma_1, M, n)$ for which

$$\int_{(0, \rho/2) \times Q_{\rho/2}(y_1, s_1)} |\nabla u|^2 \, dX dt \leq c \rho^{-2} \int_{(0, \rho) \times Q_\rho(y_1, s_1)} u^2 \, dX dt$$

(3.18)

$$\leq c^2 \int_{(0, \rho) \times Q_\rho(y_1, s_1)} |\nabla u|^2 \, dX dt.$$

Next we note from Lemma 3.3 for $(X, t) \in (0, r/2) \times Q_{r/2}(y, s)$ that

$$|u(X, t)|^2 \leq c \, x_0^{-2-n} \int_{(0, 2x_0) \times Q_{2x_0}(x, t)} u^2(Z, \tau) \, dZ d\tau \, .$$

Thus to prove the first part of Lemma 3.9 it suffices to show

$$(3.19) \qquad \rho^{-2-n} \int_{(0,\rho/4)\times Q_{\rho/4}(y_1,s_1)} u^2\, dX dt \le c\,(\rho/r)^\alpha \max_{(0,r)\times Q_r(y,s)} u$$

whenever $(y_1,s_1)\in Q_{r/2}(y,s)$ and $0<\rho<r/4$. To do this, if u is a weak solution to (1.1) for given A,B and $\rho/2\le\tau\le r/8$, we let u_0 be the weak solution to this equation with $B\equiv 0$ in $(0,\tau)\times Q_\tau(y_1,s_1)$ and $u\equiv u_0$ on the parabolic boundary of $(0,\tau)\times Q_\tau(y_1,s_1)$. If u satisfies (3.1) we define u_0 similarly, with (3.1) replacing (1.1). Existence of u_0 follows from Schauder type estimates or as in [A]. We put $w=u-u_0$ in place of ϕ in (2.12) or (3.2) and write down the resulting equations for u,u_0. Subtracting these equations and using (2.8), (2.9), we see that

$$(3.20)$$
$$\int_{(0,\tau)\times Q_\tau(y_1,s_1)} |\nabla w|^2\, dX dt \le c \int_{(0,\tau)\times Q_\tau(y_1,s_1)} |B|\,(|u||\nabla w|+|\nabla u||w|)\, dX dt.$$

From (3.20), (3.16) with $u\psi$ replaced by w,u, (3.13), and Cauchy's inequality we deduce that

$$(3.21) \qquad \int_{(0,\tau)\times Q_\tau(y_1,s_1)} |\nabla w|^2\, dX dt \le c\epsilon_1 \int_{(0,\tau)\times Q_\tau(y_1,s_1)} |\nabla u|^2\, dX dt.$$

Put

$$\Phi(f,v)=v^{-n}\int_{(0,v)\times Q_v(y_1,s_1)} |\nabla f|^2\, dX dt.$$

Then from Lemmas 3.3, 3.9 for u_0 we see when $0<v<\tau$ that

$$(3.22) \qquad \Phi(u_0,v) \le c(v/\tau)^{2\alpha}\Phi(u_0,\tau).$$

From (3.21), (3.22) we conclude for $0<v<\tau$, that

$$\Phi(u,v) \le 4(\Phi(u_0,v)+\Phi(w,v))$$

$$\le c\,(v/\tau)^{2\alpha}\Phi(u_0,\tau)+c\,(\tau/v)^n\Phi(w,\tau)$$

$$\le c\,[(v/\tau)^{2\alpha}+\epsilon_1(\tau/v)^n]\Phi(u,\tau).$$

Set $v/\tau=\theta$. Then from the above inequality it is clear that we can choose first θ and then ϵ_1 so small that

$$\Phi(u,\theta\tau) \le \tfrac12\Phi(u,\tau).$$

With θ now fixed we can iterate this inequality in the usual way starting with $\tau=r/8$ and continuing to $\tau=\rho/2$. From this iteration and (3.18) we conclude first that (3.19) is valid and second from our earlier remark that the first part of Lemma 3.9 is true when ϵ_1 is sufficiently small. The second part of this lemma follows from the first part and Harnack's inequality in a standard way (see [CFMS] or [FGS]). $\square$

Remark. Lemma 3.9 is still true if instead of assuming (3.13) in U we assume only that this inequality holds in $(0,2r)\times Q_{2r}(y,s)$. Indeed the above proof of Lemma 3.9 uses only the assumption that (3.13) holds in $(0,2r)\times Q_{2r}(y,s)$. We shall use this remark in chapter II.

Proof of Lemma 3.10. We first prove the left hand inequality in Lemma 3.10. To do this we integrate (3.1) with $v = G(r, y, s + 100r^2, \cdot)$ over $U \cap [R^n \times (-\infty, s)]$. We note that the inner normal to ∂U is $e = (1, 0, \dots, 0)$ and that $\nabla v = e|\nabla v|$ at points on ∂U with time coordinate $< s + 100r^2$. Using this remark, the divergence theorem, and (2.10) we get

(3.23)

$$\int_{\mathbb{R}^n} G(r, y, s + 100r^2, Z, s) \, dZ$$

$$= \int_{-\infty}^{s} \int_{\mathbb{R}^{n-1}} |\nabla G|^{-1} \langle \nabla G, A\nabla G \rangle (r, y, s + 100r^2, z, \tau) \, dz \, d\tau \leq 1.$$

Using (3.23) and Harnack's inequality (Lemma 3.5) we see that for (Z_1, τ_1) in the parabolic boundary of $Q_{r/2}(r, y, s + 100r^2)$ we have

(3.24) $\quad r^n \, G(r, y, s + 100r^2, Z_1, \tau_1) \leq c \int_{\mathbb{R}^n} G(r, y, s + 100r^2, Z, s) \, dZ \leq c.$

Next we observe from Lemma 3.9 with $u = 1 - \omega(\cdot, Q_r(y, s))$ and Harnack's inequality that $c\omega \geq 1$ on the parabolic boundary of $Q_{r/2}(r, y, s + 100r^2)$. From this inequality, (3.24), and the boundary maximum principle for solutions to (1.1) we conclude that

$$c^{-1} \, r^n \, G(X, t, r, y, s + r^2) \leq \omega(X, t, Q_r(y, s))$$

which is just the lefthand inequality in Lemma 3.10. To prove the righthand inequaliy, let ψ be the function in Lemma 3.9 with $\rho = 4r$ and $(y_1, s_1) = (y, s)$. Note from our assumptions on (X, t) in Lemma 3.10 that $\psi(X, t) = 0$. We put $\phi = \psi$ in (3.6a) and use (2.9), (3.13), (3.16) with $u\psi$ replaced by $G(X, t, \cdot)$, (3.18) with $u = G$, Cauchy's inequality, the Hardy - Littlewood maximal theorem, and Lemma 3.9 to deduce for χ = characteristic function of $(0, 4r) \times Q_{4r}(y, s)$ that

$$\omega(X, t, Q_{2r}(y, s)) \leq \int_{\partial U} \psi(Y, s) \, d\omega(X, t, Y, s)$$

$$= -\int_{U} [\langle A\nabla\psi, \nabla_Y G(X, t, \cdot) \rangle + G(X, t, \cdot)(\psi_s - B \nabla\psi)] \, dY \, ds$$

(3.25) $$\leq cr^{-1} \int_{(0,4r) \times Q_{4r}(y,s)} M^{(1)}(\chi \, |\nabla_Y G(X, t, Y, s)|) \, dY \, ds$$

$$\leq c \, r^{n/2} \left(\int_{(0,4r) \times Q_{4r}(y,s)} |\nabla G_Y(X, t, Y, s)|^2 \, dY \, ds \right)^{1/2}$$

$$\leq c \, r^n \, G(X, t, r, y, s - 100r^2)$$

which is the righthand inequality in Lemma 3.10. $\square$

Proof of Lemma 3.11. To prove Lemma 3.11, fix (X, t) as in this lemma and put $v(Z, \tau) = G(X, t, Z, \tau)$, for $(Z, \tau) \in U \setminus \{(X, t)\}$. Given $w \in R^{n-1}$ and $r > 0$ put

$$J_r(w) = \{Z \in R^n : 0 < z_0 \le r, \ -r \le z_i - w_i \le r \ \text{for} \ 1 \le i \le n - 1\},$$

$$S_r(w) = \partial J_r(w) \setminus \{Z : z_0 = 0\}$$

where $\partial J_r(w)$ is taken with respect to R^n. Following [FS] we first note from Harnack's inequality for (3.1) (Lemma 3.5) and Lemma 3.9 that for some $\lambda, c \ge 1$, depending only on γ_1, M, n, we have

(3.26)
$$e^{-ck^2} (z_0/\rho)^\lambda v(\rho, y_1, s_1 + 16\rho^2) \le v(Z, \tau) \le c\, e^{ck^2} (z_0/\rho)^\alpha v(k\rho, y_1, s_1 - 16\rho^2)$$

whenever $k \ge 100$, $Z \in J_{k\rho}(y_1)$, and $\tau \in (s_1 - 9\rho^2, \ s_1 + 9\rho^2)$. Here $\rho > 0$ and $(y_1, s_1) \in R^n$ with $s_1 \le t - 20\,\rho^2$. Second for $Q = (0, \rho) \times Q_\rho(y_1, s_1)$, we show as in [FS] that if

$$M_1 = \max\{v(Z, \tau) : (Z, \tau) \in \bar{Q}\}$$

$$M_2 = \max\{v(Z, \tau) : (Z, \tau) \in S_{k\rho}(y_1) \times (s_1 - 16\rho^2, s_1 + 16\rho^2)\}$$

and $M_2 \le k^\lambda M_1$, then for $k = k(\gamma_1, M, n) \ge 100$ large enough and ϵ_1 as in (3.13) small enough, we have

(3.27) $$M_3 = \max\{v(Z, s_1 + 4\rho^2) : Z \in J_{k\rho}(y_1)\} \ge \tfrac{1}{2} M_1.$$

To prove (3.27) let v_0 be the weak solution to (3.1) with $B \equiv 0$ in $\Omega = J_{k\rho}(y_1) \times (s_1 - 4\rho^2, s_1 + 4\rho^2)$, and with boundary values $v_0 = v$ on the parabolic boundary of Ω. Let $\tilde{G}$ be Green's function for (3.1) in Ω with $B \equiv 0$. We note that (3.6b) still holds if U is replaced by Ω and G by $\tilde{G}$ provided $\phi \in C_0^\infty(\Omega)$. Also (3.7) remains true with U, G replaced by $\Omega, \tilde{G}$. Using this remark and approximating $v - v_0$ by smooth functions we deduce from (3.6b) that for $(Z_1, \tau_1) \in \Omega$ we have

$$v(Z_1, \tau_1) = v_0(Z_1, \tau_1) + \int_\Omega v\, B \nabla_Z \tilde{G}(Z_1, \tau_1, Z, \tau)\, dZ d\tau = (v_0 + v_1)(Z_1, \tau_1).$$

Now in [FS] it is shown for some $c = c(\gamma_1, M, n)$ that

(3.28) $$v_0(Z, \tau) \le M_3 + e^{-k/c} M_2$$

when $(Z, \tau) \in \Omega, z \in J_{k\rho/2}(y_1)$, and $k \ge 100$. To estimate v_1 for $(Z_1, \tau_1) = (z_0^*, z^*, \tau_1) \in Q$ we write

$$v_1(Z_1, t_1) = \int_\Omega v\, B \nabla_Z \tilde{G}(Z_1, \tau_1, Z, \tau)\, dZ d\tau$$

$$= \int_{\Omega_1} \dots + \int_{\Omega_2} \dots + \int_{\Omega_3} \dots = I_1 + I_2 + I_3,$$

where

$$\Omega_1 = \Omega \cap \{(Z, \tau) : |Z - Z_1| + |\tau - \tau_1|^{1/2} < z_0^*/2\},$$

$$\Omega_2 = (\Omega \setminus \Omega_1) \cap \{(Z, \tau) : z_0 < z_0^*/2\}$$

$$\Omega_3 = \Omega \setminus (\Omega_1 \cup \Omega_2).$$

To estimate I_1, I_3 we shall need the inequality

$$(3.29) \qquad \tilde{G}(Z_1, \tau_1, Z, \tau) \leq c\,|\tau - \tau_1|^{-n/2} \exp\left[-\frac{|Z_1 - Z|^2}{c|\tau - \tau_1|}\right]$$

for some $c = c(\gamma_1, M, n) \geq 1$. (3.29) is proved in [A]. Put

$$E_i = \{(Z, \tau) : 2^{-i-1} z_0^* \leq |Z - Z_1| + |\tau - \tau_1|^{1/2} < 2^{-i} z_0^*\} \text{ for } i = 0, \pm 1, \ldots$$

$$E_i^* = E_{i+1} \cup E_i \cup E_{i-1}, i = 0, \pm 1, \ldots.$$

For $(Z_1, \tau_1) \in Q$, we note that $\tilde{G}(Z_1, \tau_1, \cdot)$ has an extension (also denoted $\tilde{G}$) to $U \cap [J_{k\rho}(y_1) \times (s_1 - 4\rho^2, \infty)]$ which is a local solution to (1.1) in this open set. In fact $\tilde{G}$ so extended is the Green's function for this set with pole at (Z_1, τ_1). Using this note, the same argument as in the proof of (3.18), and (3.29) we deduce

$$\int_{E_i \cap \Omega} |\nabla \tilde{G}|(Z_1, \tau_1, Z, \tau) dZ d\tau$$

$$\leq c(z_0^* 2^{-i})^{(n+2)/2} \left(\int_{E_i \cap \Omega} |\nabla \tilde{G}|^2 (Z_1, \tau_1, Z, \tau) dZ d\tau\right)^{1/2}$$

$$(3.30) \qquad \leq c\,(z_0^* 2^{-i})^{n/2} \left(\int_{E_i^* \cap \Omega} |\tilde{G}|^2 (Z_1, \tau_1, Z, \tau) dZ d\tau\right)^{1/2}$$

$$\leq c\,(2^i/z_0^*)^{n/2} \left(\int_{E_i^*} dZ d\tau\right)^{1/2} \leq c\, z_0^* 2^{-i}.$$

From (3.13), (3.26), and (3.30) we deduce for $(Z_1, \tau_1) \in Q$

$$|I_1| \leq c\,\epsilon_1\, e^{ck^2} (1/z_0^*)\, M_2 \int_{\Omega_1} |\nabla \tilde{G}|(Z_1, \tau_1, Z, \tau)\, dZ d\tau$$

$$(3.31) \qquad \leq c\,\epsilon_1\, e^{ck^2}\, (1/z_0^*)\, M_2 \left(\sum_{i=0}^{\infty} \int_{E_i} |\nabla \tilde{G}|(Z_1, \tau_1, Z, \tau) dZ d\tau\right)$$

$$\leq c\,\epsilon_1\, e^{ck^2}\, M_2\, (\sum_{i=0}^{\infty} 2^{-i}) \leq c\,\epsilon_1\, e^{ck^2}\, M_2.$$

To estimate I_2 we use (3.16), (3.18) for v, (3.30), and (3.26) to conclude

$$(3.32)$$
$$|I_2| \leq c\,\epsilon_1 \int_{\Omega_2} |\nabla \tilde{G}|(Z_1, \tau_1, \cdot)\, M^{(1)}(|\nabla v|)\, dZ\, d\tau$$

$$\leq c\,\epsilon_1 \left(\int_{\Omega_2} |\nabla \tilde{G}|^2\, dZ d\tau\right)^{1/2} \left(\int_{\Omega_2 \cup \Omega_1} |\nabla v|^2\, dZ\, d\tau\right)^{1/2} \leq c\,\epsilon_1\, e^{ck^2}\, M_2.$$

I_3 is estimated similarly using (3.13), (3.18), (3.26), and (3.30) for $-N \leq i \leq 0$, where N is the least positive integer greater such that $2^N > 2kn\rho/z_0^*$. We get

$$|I_3| \leq c\,\epsilon_1 \int_{\Omega_3} |\nabla \tilde{G}|(Z_1, \tau_1, \cdot)\, M^{(1)}(|\nabla v|)\, dZ d\tau$$

$$\leq c\,\epsilon_1 \left(\sum_{i=0}^{\infty} \int_{E_{-i} \cap \Omega_3} |\nabla \tilde{G}|(Z_1, \tau_1, \cdot) M^{(1)}(|\nabla v|)\, dZ d\tau \right)$$

(3.33)
$$\leq c\,\epsilon_1 \left(\sum_{i=0}^{\infty} [\int_{E_{-i} \cap \Omega_3} |\nabla \tilde{G}|^2\, dZ d\tau]^{1/2} [\int_{E_{-i} \cap \Omega} |\nabla v|^2\, dZ\, d\tau]^{1/2} \right)$$

$$\leq c\,\epsilon_1\, e^{ck^2}\, M_2\, (z_o^*/\rho)^\alpha\, (\sum_{i=0}^{N} 2^{i\alpha})$$

$$\leq c\epsilon_1\, e^{ck^2}\, M_2.$$

We conclude from (3.31)-(3.33) that

(3.34) $$\qquad\qquad |v_1|(Z_1, \tau_1) \leq c\,\epsilon_1\, e^{ck^2}\, M_2.$$

From (3.34) and (3.28) we see that first we can choose k large and then with k fixed choose ϵ_1 small enough so that if $M_2 \leq k^\gamma M_1$, then

$$M_1 \leq \tfrac{1}{2} M_1 + M_3$$

which clearly implies (3.27).

We now use (3.27) to prove Lemma 3.11. We suppose k, ϵ_1 are fixed so that Lemmas 3.9, 3.10, and (3.27) are valid. Let $(X, t), (r, y, s + 100r^2)$, be as in Lemma 3.11, and suppose $y_1 = y$, $s - 100r^2 \leq s_1 \leq s + 100r^2$. In order to avoid confusion we write $Q(\rho)$ for the above Q and put $M(\rho) = \max_{\bar{Q}(\rho)} v$. Let

$$\sigma = \max_{\{2r \leq \rho^* \leq x_0/2\}} (\rho^*)^{-\lambda} M(\rho^*)$$

and choose $\rho, 2r \leq \rho \leq x_0/2$, such that $M(\rho) = \rho^\lambda \sigma$. We consider two cases. First if $\rho \geq x_0/(2k)$, then from (3.26) and Lemma 3.10 we deduce for some $K = K(\gamma_1, M, n, k) \geq 2$ that

$$K^{-1} x_0^{-n-\lambda} \leq r^{-\lambda} v(r, y_1, s_1 + r^2) \leq c\, r^{-\lambda} v(r, y_1, s_1)$$
(3.35)
$$\leq c\,\sigma \leq K\, x_0^{-n-\lambda}.$$

Otherwise $\rho \leq x_0/(2k)$ and we have

$$M(k\rho) \leq k^\lambda\, M(\rho),$$

so (3.27) can be used to conclude

(3.36) $$\qquad\qquad M(\rho) \leq 2\, v(Z, s_1 + 4\rho^2)$$

for some $Z \in J_{k\rho}(y_1)$. From this inequality, Harnack's inequality, (3.26), and Lemma 3.9 it follows easily that for some $K = K(\gamma_1, M, n, k) \geq 2$

$$K^{-1}M(\rho) \leq 2K^{-1}v(Z, s_1 + 4\rho^2)$$

$$\leq (\rho/r)^\lambda \, v(r, y_1, s_1 + r^2) \leq c(\rho/r)^\lambda v(r, y_1, s_1)$$

$$\leq c\,M(\rho).$$

Thus in either case

$$v(r, y_1, s_1) \leq Kv(r, y_1, s_1 + r^2).$$

We first take $(y_1, s_1) = (y, s - 100r^2)$. Then we can repeat this argument with (r, y_1, s_1) replaced by $(r, y_1, s_1 + r^2)$. Doing this at most 200 times we obtain the conclusion of Lemma 3.11 . $\square$

Lemmas 3.5, 3.10, 3.11 clearly imply Lemma 3.12. The proof of Lemma 3.14 is now complete. $\square$

Finally in this section we drop the assumption that A, B are smooth when $B \not\equiv 0$. We prove

Lemma 3.37 *Let A, B satisfy (2.8)-(2.10) and (3.13). If $\epsilon_1 = \epsilon_1(\gamma_1, M, n) > 0$ is small enough, then the continuous Dirichlet problem for (1.1) has a unique solution. Moreover if ω denotes parabolic measure corresponding to (1.1), then whenever $(x, t) \in R^n$ and $Q_{2r}(y, s) \subset Q_d(x, t)$ we have for some $c = c(\gamma_1, M, n, \epsilon_1) \geq 1$,*

(α) $c\,\omega(d, x, t + 2d^2, Q_d(x, t)) \geq 1$,

(β) $c\,\omega(d, x, t + 2d^2, Q_r(y, s)) \geq \omega(d, x, t + 2d^2, Q_{2r}(y, s))$,

(γ) *If $E \subset Q_{2r}(y, s)$ is a Borel set and $\omega(2r, y, s + 8r^2, E) \geq \eta$, then $c\,\omega(d, x, t + 2d^2, E) \geq \eta\,\omega(d, x, t + 2d^2, Q_{2r}(y, s))$.*

Proof: Our proof is standard so we shall not include all the details. Let $h \in C^\infty(-\infty, \infty)$, $0 \leq h \leq 1$, with $h \equiv 0$ on $(-\infty, 1/2)$, $h \equiv 1$ on $(1, \infty)$, and $|h'| \leq 100$. Put $h_j(X, t) = h(jx_0)$, for $(X, t) \in R^{n+1}$ and $j = 1, 2, \dots$. Let $P_\lambda \geq 0$ be a parabolic approximate identity on R^{n+1} defined as in (1.6) with R^n replaced by R^{n+1}. Recall that $P \in C_0^\infty(Q_1(0, 0))$ where now $Q_1(0, 0)$ is a rectangle in R^{n+1}. Let $A \equiv A_0$, $B \equiv 0$, in $R^n \setminus U$ where A_0 is the constant matrix in (2.10) and set

$$A_j = A_0 + h_j\,P_{\delta_j}(A - A_0)$$

$$B_j = h_j\,P_{\delta_j}B, \text{ for } j = 1, 2, \dots$$

where the convolution is understood to be with respect to each entry in the above matrices and $\delta_j = (100j)^{-1}$. Clearly $A_j, B_j \in C_0^\infty(R^{n+1})$ and $A_j \equiv A_0, B_j \equiv 0$, in $\{(X, t) : x_0 \leq (2j)^{-1}\}$. Also it is easily checked that (2.8)-(2.10) and (3.13) hold for some $c = c(n)$ with γ_1, M, ϵ_1 replaced by $\gamma_1/c, cM, c\epsilon_1$. Choosing ϵ_1 still smaller if necessary, it follows that Lemma 3.14 holds for $A_j, B_j, j = 1, 2, \dots$. Finally from properties of parabolic approximate identities we have $A_j, B_j \to A, B$ pointwise almost everywhere on U as $j \to \infty$.

Let g be a bounded continuous function on R^n and let $(u_j)_1^\infty$ be the corresponding solutions to the continuous Dirichlet problem for (1.1) with A, B replaced by A_j, B_j. Thus $u_j = g$ on ∂U. Existence of $u_j, j = 1, \dots,$ follows from (2.10) and Schauder type estimates. From the interior estimates in Lemmas 3.3, 3.4, we see there exists a subsequence $(w_j)_1^\infty$ of $(u_j)_1^\infty$ with $w_j \to u \in L^2(-T, T, H^1_{\text{loc}}(G))$ as $j \to \infty$ weakly in the norm of this space. Here T, G are as in (2.11). Also thanks to Lemma 3.4 we can choose $(w_j)_1^\infty$ so that $w_j \to u$ as $j \to \infty$ uniformly on compact subsets of U. From this fact and the maximum principle for smooth solutions to (1.1) with A, B replaced by A_j, B_j, we see that $\|u\|_{L^\infty(U)} \leq \|g\|_{L^\infty(\mathbb{R}^n)}$. Let $u = g$ on ∂U.

Next we show that u is continuous on $\bar{U}$. For this purpose, given $\eta > 0$ and $(x, t) \in R^n$, choose $\delta > 0$ so that $|g(y, s) - g(x, t)| \leq \eta$ for $(y, s) \in Q_\delta(x, t)$. Then since the continuous Dirichlet problem for (1.1), corresponding to A_j, B_j, always has a unique solution there exist sequences $(\hat{w}_j), (w_j^*)$ such that for $j = 1, 2, \dots$

(i) $\hat{w}_j, w_j^*$ are solutions to the continuous Dirichlet problem for (1.1), A_j, B_j.

(ii) $w_j - g(x, t) = \hat{w}_j + w_j^*$.

(iii) $\|\hat{w}_j\|_{L^\infty(U)} \leq 2\eta$.

(iv) $w_j^* \equiv 0$ on $Q_{\delta/2}(x, t) \subset \partial U$.

Choosing certain subsequences of $(\hat{w}_j), (w_j^*)$ it follows from $(i) - (iv)$ that $u - g(x, t) = u_1 + u_2$ where u_1, u_2 are uniform limits on compact subsets of subsequences of $(\hat{w}_j), (w_j^*)$ respectively. Moreover u_1, u_2 are weak solutions to (1.1) corresponding to A, B, and $\|u_1\|_{L^\infty(U)} \leq 2\eta$. We note from (iv) that Lemma 3.9 holds for (w_j^*) in $(0, \delta/2) \times Q_{\delta/2}(x, t)$ with constants independent of j. Thus the conclusion of this lemma also holds for u_2 so that $\lim_{(Y,s) \to (x,t)} u_2(Y, s) = 0$. From this remark and the above facts we conclude that

$$\limsup_{(Y,s) \to (x,t)} |u(Y, s) - g(x, t)| \leq 2\eta.$$

Since $\eta \in R$ and $(x, t) \in R^n$ are arbitrary we see that u is a solution to the Dirichlet problem for (1.1) with boundary function g.

Uniqueness of u is a consequence of the following maximum principle.

Lemma 3.38. *Let u, v be bounded continuous local weak solutions to (1.1) in U where A, B satisfy (2.8)-(2.10). If*

$$\limsup_{(Y,s) \to (x,t)} (u - v)(Y, s) \leq 0,$$

whenever $(x, t) \in R^n$, then $u \leq v$ in U.

Proof: Given $\epsilon > 0$ we see from (2.10), continuity, and our knowledge of constant coefficient parabolic pde's, that there exists $\delta > 0$ with $u - v \leq \epsilon$ in $U \setminus G$ where $G = (\delta, 1/\delta) \times Q_{1/\delta}(0, 0)$. Also from (2.9) we deduce that B is essentially bounded in G. Now if $u - v \not\leq 0$, then for $\epsilon > 0$ small we would have $u - v > 2\epsilon$ at some point in G. We could then apply a weak maximum principle in G (valid since B is essentially bounded in this set (see [A]). Doing this we get $u \leq v + \epsilon$ in G which is a contradiction. Thus Lemma 3.38 is true. $\square$

To complete the proof of Lemma 3.37 we now define parabolic measure ω as in section 2 relative to (1.1), $A, B.$ $(\alpha), (\beta), (\gamma)$ are easily proved for smooth solutions using Lemmas 3.14, 3.12, 3.10, and 3.9. Taking a weak limit as above, it then follows that ω satisfies $(\alpha), (\beta), (\gamma)$. For completeness we sketch the proof of (γ) under the asumption that A, B are smooth in $\bar{U}$. Indeed from Lemma 3.10 and Harnack's inequality (Lemma 3.5), we deduce the existence of $\tilde{c} \geq 1$ such that $\tilde{c}\,\omega(\cdot, E) \geq r^n\, G(\cdot, 2r, y, s + 12r^2)$ on $\partial\Omega$ where $\Omega = U \setminus L$ and $L = \{(Z, \tau) \in U : |z_i - y_i| < r, |z_0 - 2r| < r, |\tau - s - 12r^2|^{1/2} \leq r, 1 \leq i \leq n - 1\}$. Using the maximum principle for solutions to (1.1), Lemma 3.10 once again, and the backward Harnack inequality (Lemma 3.11), we get (γ). The proof of Lemma 3.37 is now complete. $\square$

Remark. We note for use in chapter II that Lemma 3.38 remains valid if U is replaced by a more general region Ω. For use in chapter II we require only that this maximum principle remains valid when Ω has one of the following forms: $(a)\,(0, r) \times Q_d(x, t)$, $(b)\,U \setminus [(0, r) \times Q_d(x, t)]$, and $(c)\,U \setminus \{(0, r) \times [Q_d(x, t) \setminus Q_\rho(y, s)]\}$. Here $0 < r, d < \infty, (x, t) \in R^n$. and $Q_\rho(y, s) \subset Q_d(x, t)$.

4. PROOF OF THEOREM 2.13 IN A SPECIAL CASE

In this section we prove Theorem 2.13, except for some square function estimates, under the assumptions that

(4.1)

(a) $B \equiv 0$ and $A \in C^\infty(\bar{U})$.

(b) A satisfies (2.8)-(2.10) .

(c) $(x_0\,|\nabla A| + x_0^2\,|A_t|)(X, t) \leq \Lambda < \infty$ for a.e. $(X, t) \in U$ and the measure μ_2 with $d\mu_2(X, t) = [\,x_0\,|\nabla A|^2 + x_0^3\,|A_t|^2\,](X, t)\,dX\,dt$
   is a Carleson measure on U.

(d) $\dfrac{\partial A_{ij}}{\partial x_0} = \displaystyle\sum_{l=0}^{n-1} \langle e_l^{ij}, \frac{\partial}{\partial x_l} f_l^{ij} \rangle + g^{ij}$ in the distributional sense where

   $\displaystyle\sum_{l=0}^{n-1} [\,|e_l^{ij}| + |f_l^{ij}|\,](X, t) \leq \Lambda < \infty,$ for a.e $(X, t) \in U$ whenever
   $0 \leq i, j \leq n - 1,$

(e) e_l^{ij}, f_l^{ij} have distributional first partials and

   $d\mu_3(X, t) = [\,\displaystyle\sum_{i,j=0}^{n-1} (\sum_{l=0}^{n-1} x_0\,|\nabla e_l^{ij}|^2 + x_0^{-1}\,|f_l^{ij}|^2) + |g^{ij}|\,](X, t)\,dX\,dt$
   is a Carleson measure on U.

(f) $\|A - A_0\|_{L^\infty(U)}^2 + \|\mu_2\| + \|\mu_3\| \leq \epsilon_0^2.$

In (4.1)(f), A_0 is the constant matrix in Theorem 2.13. We note that (4.1)(a) is the only additional assumption to those in Theorem 2.13. In section 5 we shall remove (4.1)(a) and thus obtain Theorem 2.13. To this end, recall that $G(\cdot, Y, s)$

is the Green's function for (1.1) with pole at $(Y, s) \in U$ and $\omega(X, t, \cdot)$ is parabolic measure at (X, t) for (1.1). From Schauder type estimates and (2.10) we note that

$$\frac{d\omega}{dy\,ds}(X, t, y, s) = K(X, t, y, s), \ (X, t) \in U, \ (y, s) \in R^n,$$

where

(4.2)

$$T = \sup_{\{(y,s)\in R^n, r>0\}} |Q_r(y, s)| \left(\int_{Q_r(y,s)} K(r, y, s + 2r^2, z, \tau)^2 \, dz \, d\tau \right) \leq c_1 < \infty.$$

Here c_1 may depend on among other things the smoothness of A. We shall show in fact that

(4.3) $c_1 = c_1(\epsilon_0, \gamma_1, M, \Lambda, n)$.

Before we begin the proof of (4.3), we note that (4.3), the basic estimates in section 3, and a familiar " rate " type argument imply that Theorem 2.13 is valid when (4.1) holds. For completeness we give the proof of this statement. If

$$\Gamma = \sup_{(z,\tau)\in Q_r(y,s)} \frac{K(d, x, t + 2d^2, z, \tau)}{K(r, y, s + 2r^2, z, \tau)}$$

then from (4.3) we deduce for some $c = c(\gamma_1, M, n) \geq 1$ that

(4.4)
$$\int_{Q_r(y,s)} K(d, x, t + 2d^2, z, \tau)^2 \, dz \, d\tau$$
$$\leq \Gamma^2 \int_{Q_r(y,s)} K(r, y, s + 2r^2, z, \tau)^2 \, dz \, d\tau \leq c_1 \Gamma^2 \, |Q_r(y, s)|^{-1}.$$

To estimate Γ we first use Schauder regularity, the fact that $K(d, x, t + 2d^2, z, \tau) = |\nabla G|^{-1} \langle \nabla G, A\nabla G \rangle$, and (2.8) to find

$$\Gamma \leq c \sup_{(z,\tau)\in Q_r(y,s)} \lim_{z_0 \to 0} \left[\frac{G(d, x, t + 2d^2, z_0, z, \tau)}{G(r, y, s + 2r^2, z_0, z, \tau)} \right].$$

Second we observe from Harnack's inequality and Lemma 3.37 (α) with $d = r$, $(x, t) = (y, s)$ that for some $c = c(\gamma_1, M, n)$ we have

(4.5) $c\,\omega(\cdot, Q_{2r}(y, s)) \geq 1$ on $(0, r) \times Q_r(y, s)$.

Using (4.5), Harnack's inequality, and Lemma 3.9 with $u = G(\cdot, z_0, z, \tau)$ we find that

$$G(\cdot, z_0, z, \tau) \leq c\,\omega(\cdot, Q_{2r}(y, s))\,G(r, y, s + 2r^2, z_0, z, \tau) \text{ on } \partial[(0, \tfrac{5}{4}r) \times Q_{\frac{5}{4}r}(y, s)]$$

so by the boundary maximum principle for solutions to (1.1) and Lemma 3.12 we have

$$G(d, x, t + 2d^2, z_0, z, \tau) \leq c\,\omega(d, x, t + 2d^2, Q_{2r}(y, s))\,G(r, y, s + 2r^2, z_0, z, \tau)$$

$$\leq c\,\omega(d, x, t + 2d^2, Q_r(y, s))\,G(r, y, s + 2r^2, z_0, z, \tau).$$

Dividing this inequality by $G(r, y, s + 2r^2, z_0, z, \tau)$ and letting $z_0 \to 0$, we conclude for some $c = c(\gamma_1, M, n)$ that

(4.6) $\Gamma \leq c\,\omega(d, x, t + 2d^2, Q_r(y, s))$.

Using (4.6) in (4.4) it follows that

$$(4.7) \quad \begin{aligned} &|Q_r(y,s)|^{-1} \int_{Q_r(y,s)} K(d,x,t+2d^2,z,\tau)^2 \, dz \, d\tau \\ &\qquad \leq c \left(|Q_r(y,s)|^{-1} \int_{Q_r(y,s)} K(d,x,t+2d^2,z,\tau) \, dz d\tau \right)^2 \end{aligned}$$

whenever $Q_r(y,s) \subset Q_d(x,t)$. Hence to prove Theorem 2.13 when (4.1) is valid it remains to prove (4.3).

Proof of (4.3). To prove (4.3) choose $(r,y,s+2r^2) \in U$ so that

$$(4.8) \qquad T \leq 2 \, |Q_r(y,s)| \left(\int_{Q_r(y,s)} K(r,y,s+2r^2,z,\tau)^2 \, dz \, d\tau \right)$$

and for given $g \in C_0^\infty(Q_r(y,s))$, $g \geq 0$, set

$$u(X,t) = \int_{\mathbb{R}^n} K(X,t,z,\tau) g(z,\tau) dz d\tau.$$

From (3.8), (2.10), and Schauder regularity we see that u is a bounded strong solution to (1.1) in U which is continuous on $\bar{U}$ with $u = g$ on $\mathbb{R}^n$. Let u_0 be the bounded solution to the continuous Dirichlet problem in U with $L_0 u_0 = 0$ and boundary values, $u_0 \equiv g$ on $\mathbb{R}^n$. Here L_0 is defined as in (1.1) with $A(X,t) \equiv A_0$, $B(X,t) \equiv 0$, for $(X,t) \in U$.

Now from our knowledge of second order parabolic operators with constant coefficients we get that $u_0 \in C^\infty(\bar{U})$ and for some $c = c(\gamma_1, M, n)$

$$(4.9) \qquad \|Nu_0\|_{L^2(\mathbb{R}^n)} + \|Su_0\|_{L^2(\mathbb{R}^n)} \leq c \, \|g\|_{L^2(\mathbb{R}^n)}$$

where N, S are as in (2.14), (2.15). As in section 3 we let $\zeta \in C_0^\infty[(-\frac{5}{4}r, \frac{5}{4}r) \times Q_{\frac{5}{4}r}(y,s)]$ with $\zeta \equiv 1$ on $(-r,r) \times Q_r(y,s)$, $\zeta \geq 0$, and

$$r\|\nabla \zeta\|_{L^\infty(\mathbb{R}^{n+1})} + r^2 \|\tfrac{\partial}{\partial t} \zeta\|_{L^\infty(\mathbb{R}^{n+1})} \leq c < \infty$$

for some $c = c(n)$. Put $\psi = \zeta^2$. Using (3.6) with ϕ equal to a smooth extension of $u_0 \psi$ to $\mathbb{R}^{n+1}$ with compact support and writing G for $G(r,y,s+2r^2, \cdot)$, we obtain

$$(4.10)$$

$$u(r,y,s+2r^2) = \int_{\mathbb{R}^n} g \, K(r,y,s+2r^2, \cdot) \, dZ \, d\tau = -\int_U L(u_0 \psi) \, G \, dZ d\tau$$

$$= -\int_U (L\psi) \, u_0 \, G dZ \, d\tau + \int_U \langle (A + A^\tau) \nabla u_0 \, , \, \nabla \psi \rangle G \, dZ d\tau$$

$$+ \int_U \psi \, [(L_0 - L)u_0] \, G \, dZ d\tau$$

$$= I + II + III.$$

We have

$$I = -\int_U \psi_t \, u_0 \, G \, dZ \, d\tau + \sum_{i,j=0}^{n-1} \int_U \left(\tfrac{\partial}{\partial z_i} A_{ij}\right) \left(\tfrac{\partial}{\partial z_j} \psi\right) u_0 \, G \, dZ d\tau$$

(4.11)
$$+ \sum_{i,j=0}^{n-1} \int_U A_{ij} \left(\tfrac{\partial^2}{\partial z_i \partial z_j} \psi\right) u_0 \, G dZ d\tau$$

$$= I_1 + I_2 + I_3.$$

I_3 is similar to I_1, so we only treat I_1 and I_2. If $D = [0, \tfrac{5}{4} r] \times Q_{\frac{5}{4} r}(y, s)$, then from Lemma 3.10, Cauchy's inequality, and (4.9), we get

$$|I_1| \leq c r^{-2} \int_D u_0 \, G \, dZ d\tau \leq c r^{-2-n} \int_D u_0 \, dZ \, d\tau$$

$$\leq c \, r^{-(1+n)/2} \|N u_0\|_{L^2(\mathbb{R}^n)} \leq c \, r^{-(1+n)/2} \|g\|_{L^2(\mathbb{R}^n)}.$$

Also, from Cauchy's inequality, (4.1)(f), (2.16), Lemma 3.9 with $u = G$, and Lemma 3.10, we deduce

$$|I_2| \leq c \, r^{-1} \left(\int_D |\nabla A|^2 \, u_0^2 \, z_0 \, dZ d\tau\right)^{1/2} \left(\int_D z_0^{-1} \, G^2 \, dZ d\tau\right)^{1/2}$$

$$\leq c \, \epsilon_0 \, \|N u_0\|_{L^2(\mathbb{R}^n)} \, r^{-1-\alpha-n} \, (\int_D z_0^{2\alpha-1} \, dZ d\tau)^{1/2}$$

$$\leq c \, \epsilon_0 \, r^{-(1+n)/2} \, \|g\|_{L^2(\mathbb{R}^n)}$$

where as usual $c = c(\gamma_1, M, n)$. Thus, for $0 < \epsilon_0 < 1/2$,

(4.12) $$|I| \leq c \, r^{-(1+n)/2} \|g\|_{L^2(\mathbb{R}^n)}.$$

We handle II similarly.

(4.13)
$$|II| \leq c \, r^{-1} \left(\int_D |\nabla u_0|^2 \, z_0 \, dZ d\tau\right)^{1/2} \left(\int_D z_0^{-1} \, G^2 \, dZ d\tau\right)^{1/2}$$

$$\leq c \, r^{-(1+n)/2} \|S u_0\|_{L^2(\mathbb{R}^n)} < c \, r^{-(1+n)/2} \|g\|_{L^2(\mathbb{R}^n)}$$

The main term is III. We write,

$$III = \int_U \psi \, \nabla \cdot [(A - A_0) \nabla u_0] \, G \, dZ d\tau$$

$$= \int_U \langle \nabla \psi, (A_0 - A) \nabla u_0 \rangle \, G dZ d\tau + \int_U \psi \, \langle \nabla G, (A_0 - A) \nabla u_0 \rangle dZ \, d\tau$$

$$= III_1 + III_2.$$

III_1 is estimated just like II. We get for some $c = c(\gamma_1, M, n) \geq 1$, that

(4.14) $$|III_1| \leq c \, r^{-(1+n)/2} \|g\|_{L^2(\mathbb{R}^n)}.$$

For III_2 we have

$$-III_2 = \sum_{i,j=1}^{n-1} \int_U \psi \, \tfrac{\partial}{\partial z_i} G \, (A - A_0)_{ij} \, \tfrac{\partial}{\partial z_j} u_0 \, dZ d\tau$$

$$+ \sum_{i=1}^{n-1} \int_U \psi \, \tfrac{\partial}{\partial z_i} G \, (A - A_0)_{i0} \, \tfrac{\partial}{\partial z_0} u_0 \, dZ d\tau$$

(4.15)

$$+ \sum_{i=1}^{n-1} \int_U \psi \, \tfrac{\partial}{\partial z_0} G \, (A - A_0)_{0i} \, \tfrac{\partial}{\partial z_i} u_0 \, dZ d\tau$$

$$+ \int_U \psi \, \tfrac{\partial}{\partial z_0} G \, (A - A_0)_{00} \, \tfrac{\partial}{\partial z_0} u_0 \, dZ d\tau$$

$$= III_{21} + III_{22} + III_{23} + III_{24} \, .$$

To estimate III_{21} we integrate by parts in z_0 to get

$$III_{21} = -\sum_{i,j=1}^{n-1} \int_U z_0 \, \tfrac{\partial}{\partial z_0} \psi \, \tfrac{\partial}{\partial z_i} G \, (A - A_0)_{ij} \, \tfrac{\partial}{\partial z_j} u_0 \, dZ d\tau$$

$$- \sum_{i,j=1}^{n-1} \int_U z_0 \, \psi \, \big(\tfrac{\partial^2}{\partial z_0 \partial z_i} G \big) \, (A - A_0)_{ij} \, \tfrac{\partial}{\partial z_j} u_0 \, dZ d\tau$$

(4.16)

$$- \sum_{i,j=1}^{n-1} \int_U z_0 \, \psi \, \tfrac{\partial}{\partial z_i} G \, \tfrac{\partial}{\partial z_0} (A - A_0)_{ij} \, \tfrac{\partial}{\partial z_j} u_0 \, dZ d\tau$$

$$- \sum_{i,j=1}^{n-1} \int_U z_0 \, \psi \, \tfrac{\partial}{\partial z_i} G \, (A - A_0)_{ij} \, \tfrac{\partial^2}{\partial z_0 \partial z_j} u_0 \, dZ d\tau$$

$$= P_1 + P_2 + P_3 + P_4 \, .$$

Using Cauchy's inequality, (3.18) with u replaced by G, Lemma 3.10, (4.1), and (4.9) we get

$$|P_1| \leq c \, \epsilon_0 \, r^{-1/2} \left(\int_D |\nabla G|^2 \, dZ \, d\tau \right)^{1/2} \left(\int_D z_0 \, |\nabla u_0|^2 \, dZ \, d\tau \right)^{1/2}$$

$$\leq c \, \epsilon_0 \, r^{-(1+n)/2} \, \|S u_0\|_{L^2(\mathbb{R}^n)} \leq c \, \epsilon_0 \, r^{-(1+n)/2} \, \|g\|_{L^2(\mathbb{R}^n)} \, .$$

Set

$$J = \sum_{i,j=0}^{n-1} \int_U \psi \, |\tfrac{\partial^2}{\partial z_i \partial z_j} G \, |^2 \, z_0 \, dZ d\tau.$$

Then from Cauchy's inequality, (4.1)(f), and (4.9), we get

$$|P_2| \leq c \epsilon_0 \, J^{1/2} \, \|S u_0\|_{L^2(\mathbb{R}^n)} \leq c \epsilon_0 \, J^{1/2} \, \|g\|_{L^2(\mathbb{R}^n)} \, .$$

We note from (4.1)(c) that ∇G satisfies on D a system of partial differential equations similar to (1.1). Moreover Lemma 3.3 holds for this system so that if $(\hat{Z}, \hat{\tau}) \in D$, then

$$|\nabla G|(r, y, s + 2r^2, \hat{Z}, \hat{\tau}) \leq c \left(\hat{z}_0^{-(n+2)} \int_{Q_{\hat{z}_0/100}(\hat{Z}, \hat{\tau})} |\nabla G|^2 \, dZ d\tau \right)^{1/2}$$

$$\leq c \left(\hat{z}_0^{-(n+4)} \int_{Q_{\hat{z}_0/100}(\hat{Z}, \hat{\tau})} |G|^2 \, dZ d\tau \right)^{1/2}.$$

Using this inequality and Lemmas 3.10, 3.12, we conclude for $(\hat{Z}, \hat{\tau}) \in D$ that

$$|\nabla G(r, y, s + 2r^2, \tilde{Z}, \tilde{\tau})|$$

$$\leq c \hat{z}_0^{-(n+1)} \omega \left(r, y, s + 2r^2, Q_{\frac{z_0}{100}}(\hat{z}, \hat{\tau}) \right) \leq c M(\chi K(r, y, s + 2r^2, \cdot))(\hat{z}, \hat{\tau}),$$

whenever $\tilde{z}_0 = \hat{z}_0$ and $|\hat{z} - \tilde{z}| + |\hat{\tau} - \tilde{\tau}|^{1/2} \leq a \hat{z}_0$. Here χ is the characteristic function of $Q_{\frac{11}{8} r}(y, s)$, a fixed is as in the definition of a parabolic cone (above (2.14)), and M is the Hardy - Littlewood maximal function defined relative to rectangles of side length ρ, ρ^2 in the space and time variables respectively. Taking the supremum over parabolic cones it follows from the above inequality, the Hardy - Littlewood maximal theorem, and (4.8) that

(4.17)
$$\|N(|\nabla G| \psi^{1/2})\|_{L^2(\mathbb{R}^n)} \leq c \|K(r, y, s + 2r^2, \cdot) \chi\|_{L^2(\mathbb{R}^n)} \leq c \, r^{-(n+1)/2} T^{1/2}.$$

Using (4.17), (2.16), and (4.1)(f) we estimate P_3 by

$$|P_3| \leq c \left(\int_U \psi |\nabla G|^2 z_0 |\nabla A|^2 \, dZ \, d\tau \right)^{1/2} \left(\int_U \psi |\nabla u_0|^2 z_0 \, dZ \, d\tau \right)^{1/2}$$

$$\leq c \, c_0 \, r^{-(n+1)/2} T^{1/2} \|S u_0\|_{L^2(\mathbb{R}^n)} \leq c \, \epsilon_0 \, r^{-(n+1)/2} T^{1/2} \|g\|_{L^2(\mathbb{R}^n)}.$$

We can handle P_4 in the same way as P_1, P_2, P_3, after integrating by parts in $z_j, 1 \leq j \leq n-1$, to move a derivative onto G, ψ, or $A - A_0$. We get

$$|P_4| \leq c \, \epsilon_0 \, [J^{1/2} + (1 + T^{1/2}) r^{-(n+1)/2}] \|g\|_{L^2(\mathbb{R}^n)}.$$

Using these estimates for the $P's$ in (4.16) we conclude that

(4.18) $$|III_{21}| \leq c \, \epsilon_0 \, [J^{1/2} + (1 + T^{1/2}) r^{-(n+1)/2}] \|g\|_{L^2(\mathbb{R}^n)}.$$

Next we estimate III_{22} . Integrating by parts in z_0 and using the fact that $\frac{\partial}{\partial z_i}G = 0$ on ∂U for $1 \le i \le n-1$, we get

$$III_{22} = -\int_U \psi_{z_0} \sum_{i=1}^{n-1} \frac{\partial}{\partial z_i}G \, (A - A_0)_{i0} \, u_0 \, dZd\tau$$

$$-\int_U \psi \sum_{i=1}^{n-1} (\frac{\partial^2}{\partial z_i \partial z_0}G) \, (A - A_0)_{i0} \, u_0 \, dZd\tau$$

(4.19)

$$-\int_U \psi \sum_{i=1}^{n-1} \frac{\partial}{\partial z_i}G \, \frac{\partial}{\partial z_0}A_{i0} \, u_0 \, dZd\tau$$

$$= S_1 + S_2 + S_3.$$

As in the estimate for P_1 we have

$$|S_1| \le c \, r^{-1} \, \epsilon_0 \left(\int_D |\nabla G|^2 \, dZd\tau\right)^{1/2} \left(\int_D u_0^2 \, dZ \, d\tau\right)^{1/2}$$

$$\le c \, \epsilon_0 \, r^{-(n+1)/2} \, \|g\|_{L^2(\mathbb{R}^n)} \, .$$

To handle S_2 we integrate by parts in z_0 to get

$$S_2 = \int_U z_0 \frac{\partial}{\partial z_0}\psi \sum_{i=1}^{n-1} (\frac{\partial^2}{\partial z_i \partial z_0}G) \, (A - A_0)_{i0} \, u_0 \, dZd\tau$$

$$+ \int_U z_0 \, \psi \sum_{i=1}^{n-1} (\frac{\partial^3}{\partial z_i \partial z_0^2}G) \, (A - A_0)_{i0} \, u_0 \, dZd\tau$$

(4.20)

$$+ \int_U z_0 \, \psi \sum_{i=1}^{n-1} (\frac{\partial^2}{\partial z_i \partial z_0}G) \, \frac{\partial}{\partial z_0}A_{i0} \, u_0 \, dZd\tau$$

$$+ \int_U z_0 \, \psi \sum_{i=1}^{n-1} (\frac{\partial^2}{\partial z_i \partial z_0}G) \, (A - A_0)_{i0} \, \frac{\partial}{\partial z_0}u_0 \, dZd\tau$$

$$= S_{21} + S_{22} + S_{23} + S_{24} \, .$$

Now $\nabla\psi = 2\zeta\nabla\zeta = 2\psi^{1/2}\nabla\zeta$. From this fact, Cauchy's inequality, and the definition of J we deduce

$$|S_{21}| \le c\epsilon_0 \, r^{-1/2} \, J^{1/2} \, (\int_D u_0^2)^{1/2} \le c\epsilon_0 \, J^{1/2} \, \|g\|_{L^2(\mathbb{R}^n)} \, .$$

S_{24} can be handled exactly like P_2. We get

$$|S_{24}| \le c \, \epsilon_0 \, J^{1/2} \, \|g\|_{L^2(\mathbb{R}^n)} \, .$$

S_{23} can be estimated using (4.1)(c), (2.16), and (4.9) to get

$$|S_{23}| \le c \, J^{1/2} \left(\int_U \psi \, z_0 \, |\nabla A|^2 \, u_0^2 \, dZ \, d\tau\right)^{1/2} \le c \, \epsilon_0 \, J^{1/2} \, \|g\|_{L^2(\mathbb{R}^n)} \, .$$

S_{22} can be estimated in the same way as S_{21}, S_{23}, S_{24}, after integrating by parts in z_i, $1 \le i \le n-1$. Using these estimates in (4.20) we get

$$|S_2| \le c\, \epsilon_0\, J^{1/2}\, \|g\|_{L^2(\mathbb{R}^n)}\,.$$

To treat S_3 we use $(4.1)(d),(e)$ and integrate by parts. Using once again the fact that $\frac{\partial}{\partial z_i} G$ vanishes on ∂U for $1 \le i \le n-1$ we get

$$
\begin{aligned}
S_3 &= -\int_U \psi \sum_{i=1}^{n-1} \tfrac{\partial}{\partial z_i} G\,[\,(\sum_{l=0}^{n-1} \langle\, e_l^{i0}\,,\, \tfrac{\partial}{\partial z_l} f_l^{i0}\,\rangle) + g^{i0}\,]\, u_0 \, dZ\, d\tau \\
&= \int_U \sum_{i=1}^{n-1} \tfrac{\partial}{\partial z_i} G\,[\,(\sum_{l=0}^{n-1} \langle\, \tfrac{\partial}{\partial z_l}\psi\, e_l^{i0}\,,\, f_l^{i0}\,\rangle) - \psi\, g^{i0}\,]\, u_0 \, dZ\, d\tau \\
&\quad + \int_U \psi\,(\sum_{i=1}^{n-1}\sum_{l=0}^{n-1} \tfrac{\partial^2}{\partial z_i z_l} G\, \langle\, e_l^{i0}\, f_l^{i0}\,\rangle)\, u_0 \, dZ\, d\tau \\
&\quad + \int_U \psi\,(\sum_{i=1}^{n-1}\sum_{l=0}^{n-1} \tfrac{\partial}{\partial z_i} G\, \langle\, \tfrac{\partial}{\partial z_l} e_l^{i0}\,,\, f_l^{i0}\,\rangle)\, u_0 \, dZ\, d\tau \\
&\quad + \int_U \psi\,(\sum_{i=1}^{n-1}\sum_{l=0}^{n-1} \tfrac{\partial}{\partial z_i} G\, \langle\, e_l^{i0}\,,\, f_l^{i0}\,\rangle\, \tfrac{\partial}{\partial z_l} u_0)\, dZ\, d\tau \\
&= S_{31} + S_{32} + S_{33} + S_{34}\,.
\end{aligned}
$$

(4.21)

From $(4.1)(d),(e),(f)$, (4.17), and estimates similar to the ones for P_1, P_3 we see that

$$
\begin{aligned}
|S_{31}| &\le cr^{-1}\left(\int_U \psi\,|\nabla G|^2\, dZ\, d\tau\right)^{1/2}\left(\int_U \psi\, u_0^2\, dZ\, d\tau\right)^{1/2} \\
&\quad + c\left(\int_U \psi\,|\nabla G|^2 \sum_{i=1}^{n-1}|g^{i0}|dZ\, d\tau\right)^{1/2}\left(\int_U \psi\, u_0^2 \sum_{i=1}^{n-1}|g^{i0}|\, dZ\, d\tau\right)^{1/2} \\
&\le c\, r^{-(n+1)/2}\,(1 + \epsilon_0^2\, T^{1/2})\,\|g\|_{L^2(\mathbb{R}^n)}.
\end{aligned}
$$

Next we have

$$
\begin{aligned}
|S_{32}| &\le c\left(\int_U z_0\,\psi \sum_{i=1}^{n-1}\sum_{l=0}^{n-1} |\tfrac{\partial^2}{\partial z_l z_i} G|^2\, dZ\, d\tau\right)^{1/2} \\
&\quad \cdot \left(\sum_{i=1}^{n-1}\sum_{l=0}^{n-1}\int_U z_0^{-1}\,\psi\,|f_l^{i0}|^2\,|u_0|^2\, dZ\, d\tau\right)^{1/2} \le c\epsilon_0\, J^{1/2}\,\|g\|_{L^2(\mathbb{R}^n)}.
\end{aligned}
$$

Also from $(4.1)(e),(f),(2.16),(4.9)$, we deduce

$$|S_{33}| \leq c\left(\sum_{i=1}^{n-1}\sum_{l=0}^{n-1}\int_U z_0^{-1}\,\psi\,|\,f_l^{i0}\,|^2\,|\nabla G|^2\,dZ\,d\tau\right)^{1/2}$$

$$\cdot\left(\sum_{i=1}^{n-1}\sum_{l=0}^{n-1}\int_U z_0\,\psi\,|\nabla e_l^{ij}|^2\,u_0^2\,dZ\,d\tau\right)^{1/2} \leq c\,\epsilon_0\,r^{-(n+1)/2}\,T^{1/2}\,\|g\|_{L^2(\mathbb{R}^n)}.$$

Lastly,

$$|S_{34}| \leq c\left(\sum_{i=1}^{n-1}\sum_{l=0}^{n-1}\int_U z_0^{-1}\,\psi\,|\,f_l^{i0}\,|^2\,|\nabla G|^2\,dZ\,d\tau\right)^{1/2}\left(\int_U z_0\,\psi\,|\nabla u_0|^2\,dZ\,d\tau\right)^{1/2}$$

$$\leq c\epsilon_0\,r^{-(n+1)/2}\,T^{1/2}\,\|g\|_{L^2(\mathbb{R}^n)}.$$

Thus

$$|S_3| \leq c\,[r^{-(n+1)/2}(1+\epsilon_0\,T^{1/2})+\epsilon_0\,J^{1/2}]\,\|g\|_{L^2(\mathbb{R}^n)}\,.$$

Using our estimates for S_1, S_2, S_3 in (4.19) we find that

$$(4.22)\qquad |III_{22}| \leq c\,[r^{-(n+1)/2}(1+\epsilon_0\,T^{1/2})+\epsilon_0\,J^{1/2}]\,\|g\|_{L^2(\mathbb{R}^n)}.$$

We handle III_{23} in exactly the same way as III_{21}. That is we first integrate by parts in z_0 to obtain 4 terms which can be estimated in exactly the same way as $P_1 - P_4$. We get

$$(4.23)\qquad |III_{23}| \leq c\epsilon_0\,[r^{-(n+1)/2}(1+T^{1/2})+J^{1/2}]\,\|g\|_{L^2(\mathbb{R}^n)}.$$

It remains to estimate III_{24}. Integrating by parts in z_0, we have

$$III_{24} = -\int_U \frac{\partial}{\partial z_0}\psi\,\frac{\partial}{\partial z_0}G\,(A-A_0)_{00}\,u_0\,dZd\tau$$

$$-\int_U \psi\,\frac{\partial^2}{\partial z_0^2}G\,(A-A_0)_{00}\,u_0\,dZd\tau$$

$$(4.24)\qquad -\int_U \psi\frac{\partial}{\partial z_0}G\,\frac{\partial}{\partial z_0}(A-A_0)_{00}\,u_0\,dZd\tau$$

$$+\int_{\partial U}\psi\frac{\partial}{\partial z_0}G\,(A-A_0)_{00}\,g\,dZd\tau$$

$$= L_1 + L_2 + L_3 + L_4.$$

L_1 is estimated in the same way as S_1 while the boundary term L_4 is easily handled using the definition of K, Cauchy's inequality, and $(4.1)(f)$. L_3 is estimated similar to S_3, using the " integration by parts hypothesis " $(4.1)(d),(e)$. However in this case we must use a different argument to show that the boundary term corresponding to integration in z_0 vanishes, since $\frac{\partial}{\partial z_0}G \neq 0$ on ∂U. To show vanishing observe from $(4.1)(e)$ and Cauchy's inequality that there exist (δ_j) with $\lim_{j\to\infty}\delta_j = 0$ and such that

$$\int_{\mathbb{R}^n}\psi\,|f_0^{00}|(\delta_j,z,\tau)\,dzd\tau\to 0 \text{ as } j\to\infty.$$

Using this observation, $(4.1)(d)$, smoothness of G, u_0, and taking a limit as $\delta_j \to 0$, we deduce that the boundary term obtained from using $(4.1)(d)$ in L_3 and integrating by parts, vanishes. Altogether we find that

$$(4.25) \qquad |L_1| + |L_3| + |L_4| \leq c\,[r^{-(n+1)/2}(1 + \epsilon_0\, T^{1/2}) + \epsilon_0\, J^{1/2}]\,\|g\|_{L^2(\mathbb{R}^n)}.$$

To estimate L_2 we make several observations. First, observe that

$$\frac{\partial^2}{\partial x_0^2} G = A_{00}^{-1}\left[\frac{\partial}{\partial x_0}\left(A_{00}\, \frac{\partial}{\partial x_0} G\right) - \left(\frac{\partial}{\partial x_0} A_{00}\right)\frac{\partial}{\partial x_0} G\right]$$

and second that $(4.1)(c)$ holds with A_{00} replaced by A_{00}^{-1}. Using these observations we see that

$$L_2 = \int \psi\, A_{00}^{-1}\,(A_0 - A)_{00}\, \frac{\partial}{\partial z_0}\left(A_{00}\, \frac{\partial}{\partial z_0} G\right) u_0\, dZ\, d\tau$$

$$(4.26) \qquad\qquad + \int_U \psi\, A_{00}^{-1}\,(A - A_0)_{00}\, \frac{\partial}{\partial z_0} A_{00}\, \frac{\partial}{\partial z_0} G\, u_0\, dZ d\tau$$

$$= L_{21} + L_{22}.$$

L_{22} can be handled using the second observation above and $(4.1)(d)$ in essentially the same way as we handled L_3. We get the same estimate for L_{22} as for L_1, L_3, L_4 in (4.25). Using (1.1) for G we find that

$$L_{21} = \int_U \psi\, A_{00}^{-1}\,(A_0 - A)_{00}\, \frac{\partial}{\partial \tau} G\, u_0\, dZ\, d\tau$$

$$+ \sum_{i,j=1}^{n-1} \int_U \psi\, A_{00}^{-1}\,(A - A_0)_{00}\, \frac{\partial}{\partial z_i}\left(A_{ij}\, \frac{\partial}{\partial z_j} G\right) u_0\, dZ d\tau$$

$$(4.27) \qquad + \sum_{j=1}^{n-1} \int_U \psi\, A_{00}^{-1}\,(A - A_0)_{00}\, \frac{\partial}{\partial z_0}\left(A_{0j}\, \frac{\partial}{\partial z_j} G\right) u_0\, dZ\, d\tau$$

$$+ \sum_{i=1}^{n-1} \int_U \psi\, A_{00}^{-1}\,(A - A_0)_{00}\, \frac{\partial}{\partial z_i}\left(A_{i0}\, \frac{\partial}{\partial z_0} G\right) u_0\, dZ d\tau$$

$$= V_1 + V_2 + V_3 + V_4.$$

V_2, V_4 can be handled in our usual manner. For example, integrating by parts in z_0 we get

$$-V_2 = \sum_{i,j=1}^{n-1} \int_U z_0 \frac{\partial}{\partial z_0} \psi \, A_{00}^{-1} \, (A - A_0)_{00} \frac{\partial}{\partial z_i} (A_{ij} \frac{\partial}{\partial z_j} G) \, u_0 \, dZ d\tau$$

$$+ \sum_{i,j=1}^{n-1} \int_U z_0 \, \psi \frac{\partial}{\partial z_0} [A_{00}^{-1} \, (A - A_0)_{00}] \frac{\partial}{\partial z_i} (A_{ij} \frac{\partial}{\partial z_j} G) \, u_0 \, dZ d\tau$$

$$+ \sum_{i,j=1}^{n-1} \int_U z_0 \, \psi \, A_{00}^{-1} \, (A - A_0)_{00} \frac{\partial^2}{\partial z_i z_0} (A_{ij} \frac{\partial}{\partial x_j} G) \, u_0 \, dZ d\tau$$

$$+ \sum_{i,j=1}^{n-1} \int_U z_0 \, \psi \, A_{00}^{-1} \, (A - A_0)_{00} \frac{\partial}{\partial z_i} (A_{ij} \frac{\partial}{\partial z_j} G) \frac{\partial}{\partial z_0} u_0 \, dZ d\tau$$

$$= V_{21} + V_{22} + V_{23} + V_{24} \, .$$

V_{21}, V_{22}, V_{24}, are estimated using the second of the above observations, (2.16), (4.1) $(c), (f)$, (4.9), (4.17) and Cauchy's inequality, as previously. To treat V_{23} we integrate by parts in z_i, $1 \le i \le n-1$, to get integrals which can be estimated just like the other three integrals. V_4 is treated similarly. Making these estimates we find that

$$(4.28) \qquad |V_2| + |V_4| \le c\epsilon_0 \, [r^{-(n+1)/2}(1 + T^{1/2}) + J^{1/2}] \, \|g\|_{L^2(\mathbb{R}^n)}.$$

As for V_3 we have

$$V_3 = \sum_{j=1}^{n-1} \int_U \psi \, A_{00}^{-1} \, (A - A_0)_{00} \frac{\partial}{\partial z_0} A_{0j} \frac{\partial}{\partial z_j} G \, u_0 \, dZ \, d\tau$$

$$+ \sum_{j=1}^{n-1} \int_U \psi \, A_{00}^{-1} \, (A_0 - A)_{00} \, A_{0j} \frac{\partial^2}{\partial z_j z_0} G \, u_0 \, dZ d\tau.$$

The first term on the righthand side of this equality can estimated using the above observation on Carleson measures and (4.1) $(d), (e), (f)$ as in the estimate of S_3. The second term is estimated in the same way as V_2. We find that (4.28) remains valid if $|V_3|$ is added to the lefthand side of this equality. Finally integrating with respect to z_0 in the integral defining V_1 we have

$$\begin{aligned} V_1 &= \int_U z_0 \tfrac{\partial}{\partial z_0}\psi\, A_{00}^{-1}\,(A - A_0)_{00}\, \tfrac{\partial}{\partial \tau} G\, u_0\, dZ\, d\tau \\ &+ \int_U z_0\, \psi\, \tfrac{\partial}{\partial z_0}[A_{00}^{-1}\,(A - A_0)_{00}]\, \tfrac{\partial}{\partial \tau} G\, u_0\, dZ\, d\tau \\ &+ \int_U z_0\, \psi\, A_{00}^{-1}\,(A - A_0)_{00}\,(\tfrac{\partial^2}{\partial \tau \partial z_0} G)\, u_0\, dZ\, d\tau \\ &+ \int_U z_0\, \psi\, A_{00}^{-1}\,(A - A_0)_{00}\, \tfrac{\partial}{\partial \tau} G\, \tfrac{\partial}{\partial z_0} u_0\, dZ\, d\tau \\ &= V_{11} + V_{12} + V_{13} + V_{14}. \end{aligned}$$

We note that since G satisfies (1.1) we can estimate $\frac{\partial}{\partial \tau} G$ in terms of first and second partials in the space variable of G and first partials of A. Using this note and a repetition of our now well known earlier arguments we find that

$$|V_{11}| + |V_{13}| + |V_{14}| \le c\epsilon_0\, [r^{-(n+1)/2}\,(1 + T^{1/2}) + J^{1/2}]\, \|g\|_{L^2(\mathbb{R}^n)}.$$

To treat V_{12} we again integrate by parts in z_0 to see that

$$\begin{aligned} 2V_{12} &= \int_U z_0^2 \tfrac{\partial}{\partial z_0}\psi\, A_{00}^{-1}\,(A_0 - A)_{00}\,(\tfrac{\partial^2}{\partial \tau \partial z_0} G)\, u_0\, dZ\, d\tau \\ &+ \int_U z_0^2\, \psi\, \tfrac{\partial}{\partial z_0}[A_{00}^{-1}\,(A_0 - A)_{00}]\,(\tfrac{\partial^2}{\partial \tau \partial z_0} G)\, u_0\, dZ\, d\tau \\ &+ \int_U z_0^2\, \psi\, A_{00}^{-1}\,(A_0 - A)_{00}\,(\tfrac{\partial^3}{\partial \tau \partial^2 z_0} G)\, u_0\, dZ\, d\tau \\ &+ \int_U z_0^2\, \psi\, A_{00}^{-1}\,(A_0 - A)_{00}\,(\tfrac{\partial^2}{\partial \tau \partial z_0} G)\, \tfrac{\partial}{\partial z_0} u_0\, dZ\, d\tau \\ &= W_1 + W_2 + W_3 + W_4. \end{aligned}$$

To estimate these terms put

$$J_1 = \int_U z_0^3\, \psi\, |\nabla \tfrac{\partial G}{\partial \tau}|^2\, dZ d\tau\,.$$

Then using our previous arguments we deduce that

$$|W_1| + |W_2| + |W_4| \le c\epsilon_0\, [r^{-(n+1)/2}\,(1 + T^{1/2}) + J_1^{1/2}]\, \|g\|_{L^2(\mathbb{R}^n)}.$$

As for W_3 we integrate by parts in τ to get

$$\begin{aligned} W_3 &= \int_U z_0^2 \tfrac{\partial}{\partial \tau}\psi\, A_{00}^{-1}\,(A - A_0)_{00}\,(\tfrac{\partial^2}{\partial z_0^2} G)\, u_0\, dZ\, d\tau \\ &+ \int_U z_0^2\, \psi\, \tfrac{\partial}{\partial \tau}[\, A_{00}^{-1}\,(A - A_0)_{00}\,](\tfrac{\partial^2}{\partial z_0^2} G)\, u_0\, dZ\, d\tau \\ &+ \int_U z_0^2\, \psi\, A_{00}^{-1}\,(A - A_0)_{00}\, \tfrac{\partial^2}{\partial z_0^2}\, G\, \tfrac{\partial}{\partial \tau} u_0\, dZ\, d\tau. \end{aligned}$$

The first two of the above integrals on the righthand side of this equation are by now standard integrals. To estimate the third integral we need the fact that

$$\int_U \psi \, z_0^3 \, (\tfrac{\partial}{\partial \tau} u_0)^2 \, dZ d\tau \ \le \ c \, \|g\|_{L^2(\mathbb{R}^n)}$$

which is easily proven using (4.9) and well known interior estimates for derivatives of solutions to parabolic pde's with constant coeffieients. From this fact and Cauchy's inequality we get the desired estimate for the last term in the equality involving W_3. Using these estimates in the display for V_{12} we find

$$|V_{12}| \ \le \ c\epsilon_0 \, [r^{-(n+1)/2} \, (1 + T^{1/2}) + J^{1/2} + J_1^{1/2}] \, \|g\|_{L^2(\mathbb{R}^n)}.$$

Next using this inequality in the display for V_1 we see that

$$(4.29) \qquad |V_1| \ \le \ c\epsilon_0 \, [r^{-(n+1)/2} \, (1 + T^{1/2}) + J_1^{1/2} + J^{1/2}] \|g\|_{L^2(\mathbb{R}^n)}.$$

In view of (4.29),(4.28), (4.27), we see that we can first replace $|V_1|$ by $|L_{21}| = |\sum_{i=1}^4 V_i|$ in the above inequality and second in view of (4.24), (4.25), (4.26), we can replace $|L_{21}|$ by $|III_{24}|$. Using our new inequality, (4.10), (4.12), (4.13), (4.14), (4.15), (4.18), (4.22), and (4.23) we conclude that

$$(4.30)$$
$$|u(r, y, s + 2r^2)| \ \le \ c\,[r^{-(n+1)/2} \, (1 + \epsilon_0 \, T^{1/2}) + \epsilon_0 \, J_1^{1/2} + \epsilon_0 \, J^{1/2}] \, \|g\|_{L^2(\mathbb{R}^n)}.$$

We claim that

$$(4.31) \qquad\qquad\qquad J + J_1 \ \le \ c\,r^{-(n+1)} \, (1 + T)$$

where c has the same dependence as the constant in Theorem 2.13. Once (4.31) is proven we can use this inequality in (4.30) to get

$$|u(r, y, s + 2r^2)| \ \le \ c \, r^{-(n+1)/2} \, (1 + \epsilon_0 \, T^{1/2}) \, \|g\|_{L^2(\mathbb{R}^n)} \, .$$

Taking the supremum on the left hand side of this inequality over all $g \in C_0^\infty(Q_r(y, s))$ with $\|g\|_{L^2(\mathbb{R}^n)} \le 1$ and using the usual $L^2(Q_r(y, s))$ duality argument we obtain after some juggling, (4.3), (4.2). Thus it remains to prove (4.31). We prove (4.31) in section 5.

5. Proof of Theorem 2.13

We first prove (4.31) and thus complete the proof of Theorem 2.13 in the special case considered in section 4. We begin by estimating J. For $0 \leq j \leq n-1$ we write

(5.1)

$$\Theta = \int_U z_0 \, \psi \, \langle \, \nabla \tfrac{\partial}{\partial z_j} G, \, \nabla \tfrac{\partial}{\partial z_j} G \, \rangle \, dZ \, d\tau \leq c \int_U z_0 \, \psi \, \langle \, A \nabla \tfrac{\partial}{\partial z_j} G, \, \nabla \tfrac{\partial}{\partial z_j} G \, \rangle \, dZ \, d\tau$$

$$= - \int_U z_0 \, \psi \, \nabla \cdot (A \nabla \tfrac{\partial}{\partial z_j} G) \, \tfrac{\partial}{\partial z_j} G \, dZ \, d\tau$$

$$- \int_U z_0 \, \langle \nabla \psi \,, \, A \nabla \tfrac{\partial}{\partial z_j} G \, \rangle \, \tfrac{\partial}{\partial z_j} G \, dZ \, d\tau$$

$$- \int_U \psi \, \langle e_0 \,, \, A \nabla \tfrac{\partial}{\partial z_j} G \, \rangle \, \tfrac{\partial}{\partial z_j} G \, dZ \, d\tau$$

$$= \Theta_1 + \Theta_2 + \Theta_3 \,.$$

In (5.1), $e_0 = (1,0,\ldots,0)$ and $c = c(\gamma_1, M, n)$. Using Cauchy's inequality with $\epsilon's$, the fact that $\nabla \psi = 2 \psi^{1/2} \, \nabla \zeta$, (2.9), (3.18) with u replaced by G, and Lemma 3.10, we get

(5.2) $$|\Theta_2| \leq c \, r^{-(n+1)} + \tfrac{1}{4} \, |\Theta| \,.$$

As for Θ_3 we have

$$2 \, \Theta_3 = - \sum_{i=0}^{n-1} \int_U \psi \, A_{0i} \, \tfrac{\partial}{\partial z_i} \, (\tfrac{\partial}{\partial z_j} G)^2 \, dZ \, d\tau$$

$$= - \int_{\partial U} \psi \, A_{00} \, (\tfrac{\partial}{\partial z_j} G)^2 \, dz \, d\tau$$

$$+ \sum_{i=0}^{n-1} \int_U \tfrac{\partial}{\partial z_i} \psi \, A_{0i} \, (\tfrac{\partial}{\partial z_j} G)^2 \, dZ \, d\tau$$

$$+ \int_U \psi \tfrac{\partial}{\partial z_0} A_{00} \, (\tfrac{\partial}{\partial z_j} G)^2 \, dZ \, d\tau$$

$$+ \sum_{i=1}^{n-1} \int_U \psi \tfrac{\partial}{\partial z_i} A_{0i} \, (\tfrac{\partial}{\partial z_j} G)^2 dZ \, d\tau$$

$$= \Theta_{31} + \Theta_{32} + \Theta_{33} + \Theta_{34}.$$

To estimate Θ_{31} note that this term is zero unless $j = 0$ in which case we find from (2.8), (2.9) that

$$|\Theta_{31}| \leq c \, r^{-(n+1)} \, T.$$

Θ_{32} is easy to estimate using (3.18) with $u = g$ and Lemma 3.10. We get

$$|\Theta_{32}| \leq c \, r^{-(n+1)} \,.$$

To estimate Θ_{33} we use (4.1) $(d), (e), (f)$ and argue as in the estimate of S_3, L_3 to obtain

$$|\Theta_{33}| \leq cT + \tfrac{1}{16n}J.$$

To treat Θ_{34} we integrate by parts in z_0 to find that

$$\Theta_{34} = -\sum_{i=1}^{n-1} \int_U z_0 \tfrac{\partial}{\partial z_0} \psi \tfrac{\partial}{\partial z_i} A_{0i} \, (\tfrac{\partial}{\partial z_j} G)^2 dZ \, d\tau$$

$$-\sum_{i=1}^{n-1} \int_U z_0 \, \psi \tfrac{\partial^2}{\partial z_i z_0} A_{0i} \, (\tfrac{\partial}{\partial z_j} G)^2 dZ \, d\tau$$

$$-\sum_{i=1}^{n-1} \int_U z_0 \, \psi \tfrac{\partial}{\partial z_i} A_{0i} \, \tfrac{\partial}{\partial z_0}(\tfrac{\partial}{\partial z_j} G)^2 dZ \, d\tau$$

$$= \xi_1 + \xi_2 + \xi_3 \,.$$

ξ_1, ξ_3 are estimated using (4.1) $(c), (f)$, and (4.17). We get

$$|\xi_1| + |\xi_3| \leq cr^{-(n+1)}\,(1+T) + \tfrac{1}{16n}\,J \,.$$

Integrating with respect to $z_i, 1 \leq i \leq n-1$, in the integral defining ξ_2 we obtain integrals which can be handled in the same way as ξ_1, ξ_3. Putting these estimates together we see that

$$(5.3) \qquad |\Theta_3| \leq c\, r^{-(n+1)}(1+T) + \tfrac{1}{4n}\,J \,.$$

Next observe that

$$\nabla \cdot (A\nabla \tfrac{\partial}{\partial z_j} G) = \tfrac{\partial}{\partial z_j} \nabla \cdot (A\nabla G) - \nabla \cdot ((\tfrac{\partial}{\partial z_j} A)\nabla G)$$

$$= \tfrac{\partial^2}{\partial z_j \partial \tau} G - \nabla \cdot ((\tfrac{\partial}{\partial z_j} A)\nabla G) \,.$$

From this observation we find that

$$\Theta_1 = -\tfrac{1}{2} \int_U z_0 \, \psi \tfrac{\partial}{\partial \tau}(\tfrac{\partial}{\partial z_j} G)^2 \, dZ d\tau$$

$$+ \int_U z_0 \, \psi \tfrac{\partial}{\partial z_j} G \, \nabla \cdot ((\tfrac{\partial}{\partial z_j} A) \, \nabla G) \, dZ \, d\tau$$

$$= \Theta_{11} + \Theta_{12} \,.$$

To estimate Θ_{11} we integrate by parts in τ to get integrals which can be estimated just like the easy parts of Θ_2, Θ_3. As for Θ_{12} we again integrate by parts to find that

$$\Theta_{12} = - \int_U z_0 \, \tfrac{\partial}{\partial z_j} G \, \langle \nabla \psi \, , \, (\tfrac{\partial}{\partial z_j} A) \, \nabla G \rangle \, dZ \, d\tau$$

$$- \int_U z_0 \, \psi \, \langle \nabla \tfrac{\partial}{\partial z_j} G \, , \, (\tfrac{\partial}{\partial z_j} A) \, \nabla G \rangle \, dZ d\tau$$

$$- \int_U \psi \, \tfrac{\partial}{\partial z_j} G \, \langle e_0 , \, (\tfrac{\partial}{\partial z_j} A) \, \nabla G \rangle \, dZ \, d\tau$$

$$= H_1 + H_2 + H_3.$$

H_1, H_2 are easily handled using $(4.1)(c), (f)$, and (4.17). Moreover if $j = 0$ we can handle H_3 using $(4.1)(d), (f)$. To treat H_3 when $j \neq 0$, we integrate with respect to z_0 and obtain

$$H_3 = \int_U z_0 \, \tfrac{\partial}{\partial z_0} \psi \, \tfrac{\partial}{\partial z_j} G \, \langle e_0 \, , \, (\tfrac{\partial}{\partial z_j} A) \, \nabla G \rangle \, dZ \, d\tau$$

$$+ \int_U z_0 \, \psi \, \tfrac{\partial^2}{\partial z_j \partial z_0} G \, \langle e_0 \, , \, (\tfrac{\partial}{\partial z_j} A) \, \nabla G \rangle \, dZ \, d\tau$$

$$+ \int_U z_0 \, \psi \, \tfrac{\partial}{\partial z_j} G \, \langle e_0 \, , \, (\tfrac{\partial^2}{\partial z_j \partial z_0} A) \, \nabla G \rangle \, dZ \, d\tau$$

$$+ \int_U z_0 \, \psi \, \tfrac{\partial}{\partial z_j} G \, \langle e_0 \, , \, (\tfrac{\partial}{\partial z_j} A) \, \nabla \tfrac{\partial}{\partial z_0} G \rangle \, dZ \, d\tau$$

$$= H_{31} + H_{32} + H_{33} + H_{34}.$$

H_{31}, H_{32}, H_{34} are easily estimated using once again $(4.1)(c), (f)$, and (4.17). Integrating by parts with respect to z_j in the integral defining H_{33} we get integrals which can be handled either in the same way as the other H's or by using $(4.1)(d)$ as in our estimate of S_3, L_3. Using these estimates for the H's and the estimates in $(5.2), (5.3)$ in (5.1) we conclude that

$$|\Theta| \leq c \, r^{-(n+1)} (1 + T) + \tfrac{1}{2n} J$$

Summing over j, $0 \leq j \leq n - 1$, we get J on the left hand side of (5.1). Thus

$$(5.4) \qquad\qquad\qquad J \leq c \, r^{-(n+1)} (1 + T).$$

Next we consider J_1. We have

(5.5)
$$J_1 = \int_U z_0^3 \psi \langle \nabla \tfrac{\partial}{\partial \tau} G, \nabla \tfrac{\partial}{\partial \tau} G \rangle \, dZ \, d\tau \leq c \int_U z_0^3 \psi \langle \nabla \tfrac{\partial}{\partial \tau} G, A \nabla \tfrac{\partial}{\partial \tau} G \rangle \, dZ \, d\tau$$

$$= - \int_U z_0^3 \psi \tfrac{\partial}{\partial \tau} G \, \nabla \cdot (A \nabla \tfrac{\partial}{\partial \tau} G) \, dZ \, d\tau$$

$$- \int_U z_0^3 \tfrac{\partial}{\partial \tau} G \langle \nabla \psi, A \nabla \tfrac{\partial}{\partial \tau} G \rangle \, dZ \, d\tau$$

$$- 3 \int_U z_0^2 \psi \tfrac{\partial G}{\partial \tau} \langle e_0, A \nabla \tfrac{\partial}{\partial \tau} G \rangle \, dZ \, d\tau$$

$$= \Phi_1 + \Phi_2 + \Phi_3.$$

As in the estimate for Θ_2 we find

$$|\Phi_2| + |\Phi_3| \leq c \left(\int_U z_0 \psi \, (\tfrac{\partial}{\partial \tau} G)^2 \, dZ d\tau \right)^{1/2} J_1^{1/2}$$

$$\leq c r^{-(n+1)}(1+T) + cJ + \tfrac{1}{4} J_1.$$

Here we have used (1.1), (4.1) (c) to estimate $\tfrac{\partial G}{\partial \tau}$ in terms of the first and second partials in the space variable of G. To treat Φ_1 we observe that

$$\nabla \cdot (A \nabla \tfrac{\partial}{\partial \tau} G) = \tfrac{\partial}{\partial \tau} \nabla \cdot (A \nabla G) - \nabla \cdot (\tfrac{\partial}{\partial \tau} A \nabla G)$$

$$= \tfrac{\partial^2}{\partial \tau^2} G - \nabla \cdot (\tfrac{\partial}{\partial \tau} A \nabla G).$$

From this observation we find

$$\Phi_1 = -\tfrac{1}{2} \int_U z_0^3 \psi \, \tfrac{\partial}{\partial \tau} (\tfrac{\partial}{\partial \tau} G)^2 \, dZ d\tau$$

$$+ \int_U z_0^3 \psi \tfrac{\partial}{\partial \tau} G \, \nabla \cdot ((\tfrac{\partial}{\partial \tau} A) \nabla G) dZ \, d\tau$$

$$= \Phi_{11} + \Phi_{12}.$$

To estimate Φ_{11} we integrate by parts in τ to obtain an integral which is estimated similar to Φ_2, Φ_3. As for Φ_{12} we again integrate by parts to find that

$$\Phi_{12} = -\int_U z_0^3 \tfrac{\partial}{\partial \tau} G \langle \nabla \psi, (\tfrac{\partial}{\partial \tau} A) \nabla G \rangle \, dZ d\tau$$

$$- \int_U z_0^3 \psi \langle \nabla \tfrac{\partial}{\partial \tau} G, \tfrac{\partial}{\partial \tau} A \nabla G \rangle \, dZ d\tau$$

$$- 3 \int_U z_0^2 \psi \tfrac{\partial}{\partial \tau} G \langle e_0, (\tfrac{\partial}{\partial \tau} A) \nabla G \rangle \, dZ \, d\tau.$$

These integrals are easily handled as in the estimate for Φ_2, Φ_3. We get that

$$|\Phi_{12}| \leq c r^{-(n+1)}(1+T) + cJ + \tfrac{1}{4} J_1.$$

Combining this with our earlier estimates we deduce from (5.5) that

$$|J_1| \leq cr^{-(n+1)}(1+T) + cJ.$$

From this inequality and (5.4) we conclude first that (4.31) is valid and second from our earlier remarks that Theorem 2.13 is true in the special case we have been considering. $\square$

To continue the proof of Theorem 2.13 we next consider the case when

(a) A satisfies $(4.1)(a) - (f)$.

(b) $B \not\equiv 0$ and $B \in C^{\infty}(\bar{U})$.

(5.6)

(c) $x_0 |B| (X,t) \leq \epsilon_0 < \infty$ for a.e. $(X,t) \in U$ and the measure μ_1 defined by $d\mu_1(X,t) = x_0 |B|^2(X,t)dX\,dt$ is a Carleson measure on U with $\|\mu_1\| \leq \epsilon_0^2$.

To prove Theorem 2.13 in this case we use the same strategy as in section 4. We assume that $\epsilon_0 < \epsilon_1$ in Lemma 3.14. Let $\omega(X,t,\cdot)$ be parabolic measure at (X,t) corresponding to (1.1) with $B \not\equiv 0$. As in section 4 put

$$\frac{d\omega}{dy\,ds}(X,t,y,s) = K(X,t,y,s), \; (X,t) \in U, \; (y,s) \in R^n,$$

and set

$$T = \sup_{\{(y,s) \in \mathbb{R}^n, r>0\}} |Q_r(y,s)| \left(\int_{Q_r(y,s)} K(r,y,s+2r^2,z,\tau)^2 \, dz\,d\tau \right).$$

Using Lemma 3.14 and arguing as in section 4 we see that to prove Theorem 2.13 under assumption (5.6) it suffices to show that

$$T \leq c_1 = c_1(\epsilon_0, \gamma_1, M, \Lambda, n) < \infty.$$

To prove this inequality we shall need an analogue of (4.9) for solutions to (1.1) under assumption (4.1). That is given $g \geq 0 \in C_0^{\infty}(Q_r(y,s))$, let

$$u(X,t) = \int_U \hat{K}(X,t,z,\tau)g(z,\tau)dz\,d\tau.$$

where $\hat{K}$ is defined as in section 4 relative to (1.1) with $B \equiv 0$. Recall that $u \in C^{\infty}(\bar{U})$, u is a bounded strong solution to (1.1) on U with $B \equiv 0$ and $u = g$ on R^n. Using our work in section 4 we shall prove for some c having the same dependence as c_1 above that

(5.7) $\|Nu\|_{L^2(\mathbb{R}^n)} + \|Su\|_{L^2(\mathbb{R}^n)} \leq c\|g\|_{L^2(\mathbb{R}^n)}$

where N, S are as in (2.14), (2.15) with $a = 1$. Once (5.7) is proved we can proceed as in section 4 to get Theorem 2.13 under assumption (5.6). To prove (5.7) we first claim that whenever $(x,t) \in R^n$ and $d > 0$, we have $\hat{K}(d,x,t+2d^2,\cdot) \in \beta_p(Q_d(x,t))$ for some $p = p(\epsilon_0, \gamma_1, M, \Lambda, n) > 2$ with

(5.8) $\|\hat{K}(d,x,t+2d^2,\cdot)\|_{\beta_p(Q_d(x,t))} \leq c(\epsilon_0, \gamma_1, M, \Lambda, n) < \infty.$

To prove this claim observe that (5.8) with p replaced by 2 is just Theorem 2.13 in the special case proved in section 4. Now (5.8) with $p = 2$ implies (5.8) for $p > 2$. (see [CF]). Next given $(x,t) \in R^n$ and $d > 0$ let $(d,y,s) \in U$ with $|y - x| + |s -$

$t|^{1/2} \leq d$. Let $\phi_j \geq 0 \in C_0^\infty(R^n)$ with $\phi_j \equiv 1$ on $Q_{2^{j+1}d}(x,t) \setminus Q_{2^j d}(x,t)$ and supp $\phi_j \subset Q_{2^{j+2}d}(x,t) \setminus Q_{2^{j-1}d}(x,t)$ for $j = 1, 2, \ldots,$. We have

(5.9)
$$u(d,y,s) \leq \int_{Q_{4d}(x,t)} \hat{K}(d,y,s,\cdot)\, g\, dzd\tau + \sum_{j=1}^\infty \int_{\mathbb{R}^n} \hat{K}(d,y,s,\cdot)\, \phi_j\, g\, dzd\tau$$

$$= \big(\sum_{j=0}^\infty u_j\big)(d,y,s)$$

where

$$u_0(Y_1,s_1) = \int_{Q_{4d}(x,t)} \hat{K}(Y_1,s_1,\cdot)\, g\, dzd\tau$$

$$u_j(Y_1,s_1) = \int_{\mathbb{R}^n} \hat{K}(Y_1,s_1,\cdot)\, \phi_j\, g\, dzd\tau,\ j = 1, \ldots,\ .$$

We note that u_j is a solution to (1.1) with $B \equiv 0$ and u_j has continuous boundary values with $u_j \equiv 0$ on $Q_{2^{j-1}d}(x,t)$ for $j = 1, 2, \ldots$. Let $q = p/(p-1)$ be the conjugate exponent to p in (5.8). Then from this remark, Lemma 3.9, Hölder's inequality, (5.8), and Lemma 3.11 we see for $j \geq 1$ that

$$u_j(d,y,s) \leq c\, 2^{-j\alpha}\, u_j(2^{j+1}d, y, s + 24^{j+1}d^2)$$

$$\leq c\, 2^{-j\alpha}\, (Mg^q)^{1/q}(d,x,t+2d^2)\, |Q_{2^j d}(x,t)|^{1/q}$$

$$\cdot \left[\int_{Q_{2^{j+2}d}(x,t)} K(2^{j+1}d, y, s + 24^{j+1}d^2, \cdot)^p\, dzd\tau\right]^{1/p}$$

$$\leq c\, 2^{-j\alpha}\, (Mg^q)^{1/q}\, (d,x,t+2d^2)\,.$$

Here M denotes the Hardy - Littlewood maximal function defined with respect to rectangles of length ρ in the space variable and ρ^2 in the time variable. This inequality also holds when $j = 0$ as we see from Hölder's inequality and (5.8). Using these estimates for $u_j(d,y,s)$ in (5.9) we conclude that

$$u(d,y,s) \leq c\, (Mg^q)^{1/q}\,.$$

Since $1 < q < 2$, and $|x - y| + |s - t|^{1/2} \leq d$ it follows first that

$$Nu(x,t) \leq c\, (Mg^q)^{1/q}(x,t),$$

and second from the Hardy - Littlewood maximal theorem that (5.7) holds for Nu.

To prove (5.7) for Su we argue as in the proof of (4.31). We have for $c_+ \geq 1$ large enough that

(5.10)

$$c_+^{-2} \|Su\|_{L^2(\mathbb{R}^n)}^2 \leq c_+^{-1} \int_U z_0 \langle \nabla u, \nabla u \rangle \, dZ \, d\tau \leq \int_U z_0 \langle A\nabla u, \nabla u \rangle \, dZ \, d\tau$$

$$= -\int_U z_0 \nabla \cdot (A\nabla u) \, u \, dZ \, d\tau - \int_U \langle A\nabla u, e_0 \rangle u \, dZ \, d\tau$$

$$= -\frac{1}{2} \int_U z_0 \frac{\partial}{\partial \tau} u^2 \, dZ \, d\tau - \int_U \langle A\nabla u, e_0 \rangle u \, dZ \, d\tau$$

$$= 0 + I.$$

In (5.10) all integrations can be justified using Schauder regularity, (2.10), and our knowledge of constant coefficient second order parabolic pde's . To continue the estimate we note that

$$2I = -\sum_{i=0}^{n-1} \int_U A_{0i} \frac{\partial}{\partial z_i} (u)^2 \, dZ \, d\tau$$

$$= -\int_{\partial U} A_{00} g^2 \, dz d\tau$$

(5.11)

$$+ \int_U \frac{\partial}{\partial z_0} A_{00} u^2 \, dZ d\tau$$

$$+ \sum_{i=1}^{n-1} \int_U \frac{\partial}{\partial z_i} A_{0i} u^2 \, dZ \, d\tau$$

$$= I_1 + I_2 + I_3$$

From (2.9) we deduce that

$$|I_1| \leq c \|g\|_{L^2(\mathbb{R}^n)}^2 \, .$$

To handle I_2 we use the " integration by parts " hypothesis (4.1) (d) and (5.7) for Nu as in the estimates of L_3, S_3 to get

$$|I_2| \leq c\|g\|_{L^2(\mathbb{R}^n)}^2 + \frac{1}{4c_+^2} \|Su\|_{L^2(\mathbb{R}^n)}^2$$

where c_+ is as in (5.10). As for I_3 we integrate by parts in z_0 to find that

$$I_3 = -\sum_{i=1}^{n-1} \int_U z_0 \frac{\partial^2}{\partial z_i z_0} A_{0i} u^2 \, dZ \, d\tau$$

$$- \sum_{i=1}^{n-1} \int_U z_0 \frac{\partial}{\partial z_i} A_{0i} \frac{\partial}{\partial z_0} u^2 \, dZ \, d\tau$$

$$= I_{31} + I_{32}$$

Using $(4.1)(c), (e)$ we deduce that

$$|I_{32}| \leq c\|g\|^2_{L^2(\mathbb{R}^n)} + \frac{1}{4\,c_+^2}\|Su\|^2_{L^2(\mathbb{R}^n)}.$$

Integrating by parts with respect to z_i, $1 \leq i \leq n-1$ in the integrals defining I_{31} we get integrals which can be estimated in the same way as I_{32}. Putting these estimates together we find that

$$|I_3| \leq c\|g\|^2_{L^2(\mathbb{R}^n)} + \frac{1}{2\,c_+^2}\|Su\|^2_{L^2(\mathbb{R}^n)}.$$

Using this estimate as well as our earlier estimates for I_1, I_2 in (5.11), (5.10) we see that (5.7) holds for Su.

We now prove Theorem 2.13 when (5.6) holds. Let g be as above and let u_1 be the solution to the Dirichlet problem for (1.1) with $B \not\equiv 0$ and $u_1 \equiv g$ on $\mathbb{R}^n$. Let G denote the Green's function for (1.1) with $B \not\equiv 0$. Also let L, L_1 denote the operators in (1.1) with $B \equiv 0$, $B \not\equiv 0$, respectively. Proceeding as in section 4 we write G for $G(r, y, s + 2r^2, \cdot)$ and use (3.6) to obtain

(5.12)

$$u_1(r, y, s + 2r^2) = \int_{\mathbb{R}^n} g\, K(r, y, s + 2r^2, \cdot)\, dZ\, d\tau = -\int_U L_1(u\psi)\, G\, dZ d\tau$$

$$= -\int_U (L_1\psi)\, u\, G\, dZ\, d\tau + \int_U \langle (A + A^\tau)\,\nabla u, \nabla \psi \rangle\, G\, dZ d\tau$$

$$+ \int_U \psi\, [(L - L_1)u]\, G\, dZ d\tau$$

$$= T_1 + T_2 + T_3.$$

Using Lemma 3.10 (permissible by Lemma 3.14) and (5.7) we deduce

(5.13)
$$|T_1| \leq c\, r^{-(1+n)/2} \|g\|_{L^2(\mathbb{R}^n)}.$$

We can handle T_2 similarly thanks to (5.7),

(5.14)
$$|T_2| \leq c\, r^{-1} \|Su\|_{L^2(\mathbb{R}^n)} \left(\int_{Q_{2r}(y,s)} z_0^{-1}\, \psi G^2\, dZ d\tau \right)^{1/2}$$

$$\leq c\, r^{-(1+n)/2} \|g\|_{L^2(\mathbb{R}^n)}.$$

Next

$$|T_3| = |\int_U \psi\, B\, \nabla u\, G\, dZ d\tau|$$

(5.15)
$$\leq c\|Su\|_{L^2(\mathbb{R}^n)} \left(|\int_U \psi\, z_0|B|^2\, (z_0^{-1}G)^2\, dZ d\tau \right)^{1/2}$$

$$= c\|Su\|_{L^2(\mathbb{R}^n)}\, T_4.$$

From Lemma 3.10 and Harnack's inequality we see as in section 4 that

$$N(\psi^{1/2}\, z_0^{-1}\, G) \leq c\, M\left[K(r, y, s + 2r^2, \cdot)\chi \right]$$

where χ is the characteristic function of $Q_{2r}(y,s)$ and M is the Hardy - Littlewood maximal function taken with respect to rectangles. Using the above inequality, (2.16), and (5.6)(c) we get

$$T_4 \leq c\,\epsilon_0\, r^{-(n+1)/2}\, T^{1/2}\,.$$

Putting this inequality in (5.15) and using (5.7) we see that

$$|T_3| \leq c\,\epsilon_0\, r^{-(n+1)/2}\, T^{1/2}\, \|g\|_{L^2(\mathbb{R}^n)}.$$

The above inequality and (5.13), (5.14) imply

$$u(r,y,s+2r^2) \leq cr^{-(n+1)/2}\,(\,1+\epsilon_0\,T^{1/2}\,)\,\|g\|_{L^2(\mathbb{R}^n)}.$$

Taking the supremum of the lefthandside of this inequality over $g \in C_0^\infty(Q_r(y,s))$ and using $L^2(Q_r(y,s))$ duality we get $T \leq c_1 < \infty$. The proof of Theorem 2.13 is now complete when (5.6) holds. $\square$

Finally we remove the assumption $A, B \in C^\infty(\bar{U})$ in (5.6). Let $h \in C^\infty(-\infty,\infty)$, $0 \leq h \leq 1$, with $h \equiv 0$ on $(-\infty,1/2)$, $h \equiv 1$ on $(1,\infty)$, and $|h'| \leq 100$. Put $h_j(X,t) = h(jx_0)$, for $(X,t) \in \mathbb{R}^{n+1}$ and $j = 1,2,\dots$. Let P_λ be a parabolic approximate identity on $\mathbb{R}^{n+1}$ defined as in (1.6) with $\mathbb{R}^n$ replaced by $\mathbb{R}^{n+1}$. Recall that $P \in C_0^\infty(Q_1(0,0))$, $(Q_1(0,0) = $ rectangle in $\mathbb{R}^{n+1}$ $)$ and $P_\lambda \psi$ denotes convolution of P_λ with ψ. Now suppose that A, B satisfy the conditions of Theorem 2.13 so that not necessarily are $A, B \in C^\infty(\bar{U})$. Extend A, B to $\mathbb{R}^{n+1}$ by setting $A \equiv A_0$, $B \equiv 0$ in the complement of U. We assume as we may that the constant matrix A_0 in Theorem 2.13 equals the constant matrix in (2.10), since otherwise we can replace A_0 by this matrix. Next we put

$$A^j = A_0 + h_j\, P_{\delta_j}(A - A_0)$$
(5.16)
$$B^j = h_j\, P_{\delta_j} B, \text{ for j} = 1, 2, \dots,$$

where the convolution is understood to be with respect to each entry in the above matrices and $0 < \delta_j \leq (100j)^{-1}$. Clearly $A^j, B^j \in C_0^\infty(\mathbb{R}^{n+1})$ and $A^j \equiv A_0$, $B^j \equiv 0$, in $\{(X,t) : x_0 \leq (2j)^{-1}\}$. Now for fixed j and $0 < \delta_j \leq (100\,j)^{-1}$, sufficiently small, we can verify (2.8) -(2.10), (4.1)$(b) - (f)$, and (5.6)(c) using the assumptions in Theorem 2.13 for A, B, and well known convergence properties of approximate identities in Sobolev spaces. For example if

$$E^j_{kl} = h_j\, \tfrac{\partial}{\partial x_0} P_{\delta_j}[(A - A_0)_{kl}] - h_j\, \tfrac{\partial}{\partial x_0}(A - A_0)_{kl}\,,$$

then E^j_{kl} tends to zero pointwise and in $L^1(\mathbb{R}^{n+1})$ as $\delta_j \to 0$ thanks to (2.10) and (4.1)(c). Also E^j_{kl} has support in $Q_{2\rho}(0,0) \cap \{(X,t) : x_0 \geq (2j)^{-1}\}$. Choose δ_j so small that

$$\|E^j_{kl}\|_{L^1(\mathbb{R}^{n+1})} \leq \epsilon_0^3/j^{n+1}\,.$$

From this inequality we see that the measure $|E^j_{kl}|(X,t)\,dX\,dt$ is a Carleson measure with norm $\leq c(n)\,\epsilon_0^3$. Next we note that

$$\tfrac{\partial}{\partial x_0} A^j_{kl} = E^j_{kl} + (\tfrac{\partial}{\partial x_0} h_j)\, P_{\delta_j}(A - A_0) + h_j\, \tfrac{\partial}{\partial x_0} A_{kl}$$

$$= E^j_{kl} + F^j_{kl} + G^j_{kl}\,.$$

From $(4.1)(f)$ we find that the measure $|F_{kl}^j|(X, t)\, dX dt$ is a Carleson measure on U with norm $\leq c(n)\, \epsilon_0$. Using $(4.1)(d)$ we see that

$$G_{kl}^j = \sum_{m=0}^{n-1} \langle e_m^{kl}, \tfrac{\partial}{\partial x_m}(f_m^{kl}\, h_j)\rangle + g^{kl} - \langle e_0^{kl}, f_0^{kl}\rangle \tfrac{\partial}{\partial x_0} h_j$$

Also from $(4.1)(d)$ we get that the measure $|\langle e_0^{kl}, f_0^{kl}\rangle \tfrac{\partial}{\partial x_0} h_j|\, dX dt$ is a Carleson measure with norm $\leq c(n)\, \epsilon_0$. From the above display for G_{kl}^j and our observations on Carleson measures we deduce that $(4.1)(e), (f)$ holds for A_{kl}^j, $0 \leq k, l \leq n-1$, with f_m^{kl} replaced by $f_m^{kl}\, h_j$ and g^{kl} replaced by

$$E_{kl}^j + F_{kl}^j + g^{kl} - \langle e_0^{kl}, f_0^{kl}\rangle \tfrac{\partial}{\partial x_0} h_j\,.$$

Hence $A_j, B_j, j = 1, \ldots,$ satisfy (5.6) with ϵ_0 replaced by $c(n)\epsilon_0^{1/2}$. Choosing ϵ_0 still smaller if necessary we can use Theorem 2.13 for smooth coefficients to conclude that $\frac{d\omega_j}{dyds}(d, x, t + 2d^2, \cdot) \in \beta_2(Q_d(x, t))$ with constants independent of j. Now we can choose an $L^2(Q_d(x, t))$ subsequence of $(\frac{d\omega_j}{dyds}(d, x, t + 2d^2, \cdot))_1^\infty$ restricted to $Q_d(x, t)$ which converges weakly to k in this space. From Lemma 3.37 and weak convergence it is easily seen that $k = \frac{d\omega}{dyds}(d, x, t + 2d^2, \cdot)$ where ω is parabolic measure defined relative to $(1.1), A, B$. From weak convergence it follows that $\frac{d\omega}{dyds}(d, x, t + 2d^2, \cdot) \in \beta_2(Q_d(x, t))$ with reverse Hölder constant $c^* = c^*(\epsilon_0, \gamma_1, M, \Lambda, n)$. The proof of Theorem 2.13 is now complete. $\square$

Remark We note that Lemma 3.14 remains valid for sufficiently small $\epsilon_1 > 0$ if (3.13) is replaced by the assumption that $\|\mu_1\| \leq \epsilon_1$. The proof of this version of Lemma 3.14 can be obtained by copying the old proof verbatim except that whenever (3.16) is used in the old proof one uses instead the fact that μ_1 is a Carleson measure and makes $L^2(R^n)$ estimates for certain nontangential maximal functions. For example if $x_0^{-1/2} u$ denotes the function $(X, t) \rightarrow x_0^{-1/2} u(X, t)$, where u is as in Lemma 3.9, then one can estimate I_4 in this lemma by using in place of (3.16) the fact that

$$N(x_0^{-1/2} u)(X, t) \leq c M f(x, t)\,.$$

Here

$$f(x, t) = \left(\int_0^{2r} |\nabla u|^2(z_0, x, t)\, dz_0\right)^{1/2}\,.$$

The smallness assumptions in Theorem 2.13 can be weakened. Indeed suppose that A has distributional partials in X, t satisfying

(5.17)
$$x_0\, |\nabla A|(X, t) \leq \Lambda$$

$$x_0^2\, |A_t|\, (X, t) \leq \epsilon_0$$

for a.e $(X, t) \in U$ and if

$$d\mu_{21}(X, t) = x_0\, |\nabla A|^2\, (X, t)\, dX dt$$

$$d\mu_{22}(X, t) = x_0^3\, |A_t|^2\, (X, t)\, dX dt$$

then μ_{21}, μ_{22} are Carleson measures on U with

$$\|\mu_{21}\| \leq \Lambda^2$$

(5.18)

$$\|\mu_{22}\| \leq \epsilon_0^2.$$

We also assume that e_l^{ij}, f_l^{ij}, g^{ij} have the properties preceding (2.5) when either $i = 0, 0 \leq j, l \leq n-1$, or $j = 0, 0 \leq i, l \leq n-1$, and

$$\sum_{i,j} \sum_{l=0}^{n-1} |e_l^{ij}|(X,t) \leq \epsilon_0$$

(5.19)

$$\sum_{i,j} \sum_{l=0}^{n-1} |f_l^{ij}|(X,t) \leq \Lambda$$

for almost every (X,t). Moreover if

$$d\mu_{31}(X,t) = [\sum_{i,j} (\sum_{l=0}^{n-1} x_0 |\nabla e_l^{ij}|^2) + |g^{ij}|] (X,t)\, dX dt$$

$$d\mu_{32}(X,t) = \sum_{i,j} \sum_{l=0}^{n-1} x_0^{-1} |f_l^{ij}|^2 (X,t)\, dX dt,$$

then μ_{31}, μ_{32} are Carleson measures on U with

$$\|\mu_{31}\| \leq \epsilon_0$$

(5.20)

$$\|\mu_{32}\| \leq \Lambda.$$

The conclusion of Theorem 2.13 is still true if $\epsilon_0 = \epsilon_0(\gamma_1, M, \Lambda, n) > 0$ is sufficiently small and

(+) $\qquad \|\mu_1\| \leq \epsilon_0$ where μ_1 is as in (2.1),

(++) $\qquad A, B$ satisfy (2.8)-(2.10),

(+++) $\qquad$ (5.17)-(5.20) are valid.

6. REFERENCES

[A] D Aronson, *Non negative solutions of linear parabolic equations*, Ann. Scuola Norm. Sup. Pisa **22** (1968), 607-694.

[CF] R. Coifman and C. Fefferman, *Weighted norm inequalities for maximal functions and singular integrals*, Studia Math. **51** (1974), 241-250.

[CFMS] L. Caffarelli, E. Fabes, S. Mortola, and S. Salsa, *Boundary behavior of nonnegative solutions of elliptic operators in divergence form*, Indiana Univ Math. J. **30** (1981), 621-640.

[D] B. Dahlberg, *On estimates of harmonic measure*, Arch. Rational Mech. Anal. **65** (1977), 275-288.

[D1] B. Dahlberg, *On the absolute continuity of elliptic measures*, Am. J. Math. **108** (1986), 1119-1138.

[DG] E. De Giorgi, *Sulla differenziabilita e analiticita delle estremali degli integrali multipli regolari*, Mem. Accad. Sci. Torino **3** (1957), 25-42.

[F] A. Friedman, *Partial differential Equations of Parabolic Type*, Prentice-Hall, Englewood Cliffs, NJ 1964; reprinted by Robert E. Krieger, Malabar, FL 1983.

[FGS] E. Fabes, Garofalo, and Salsa, *A backward Harnack inequality and Fatou theorem for nonnegaive solutions of parabolic equations*, Illinois J. of Math. **30** (1986), 536-565.

[FJK] E. Fabes, D. Jerison, and C. Kenig, *Necesary and sufficient conditions for absolute continuity of elliptic harmonic measure*, Ann. Math. **119** (1984), 121-141.

[FKP] B. Fefferman, C. Kenig, and J. Pipher, *The theory of weights and the Dirichlet problem for elliptic equations*, Ann. Math. **134** (1991), 65-124.

[FS] E. Fabes and M. Safonov, *Behaviour near the boundary of positive solutions of second order parabolic equations*, J. Fourier Anal. Appl. **3** (1997), 871-882.

[HL] S. Hofmann and J. Lewis, *Solvability and representation by caloric layer potentials in time-varying domains*, Ann. Math. **144** (1996), 349-420.

[KKPT] C.Kenig, H. Koch, J. Pipher, and T. Toro, *A new approach to absolute continuity of elliptic measure,with applications to nonsymmetric equations*, submitted.

[KW] R. Kaufmann and J.M. Wu, *Parabolic measure on domains of class $Lip_{1/2}$*, Compositio Mathematica **65** (1988), 201-207.

[LM] J. Lewis and M. A. Murray, *The method of layer potentials for the heat equation in time-varying domains*, Memoirs of the AMS. **545** (1995), 1-157.

[M] J. Moser, *A Harnack inequality for parabolic differential equations*, Comm. on Pure and Applied Math. **17** (1964), 101-134.

[M1] J. Moser, *On a pointwise estimate for parabolic differential equations*, Comm. on Pure and Appl. Math. **24** (1971), 727-740.

[N] J. Nash, *Continuity of solutions of parabolic and elliptic equations*, American J. Math. **80** (1958), 931-954.

[St] E. Stein, *Singular integrals and differentiability properties of functions*, Princeton University Press, Princeton, 1970.

[VV] G. Verchota and A. Vogel, *Nonsymmetric systems on nonsmooth domains*, TAMS **349** (1997), no.11, 4501-4535.

CHAPTER II
ABSOLUTE CONTINUITY AND THE L^p DIRICHLET PROBLEM : PART 1

1. Introduction

Recall that in chapter I we considered weak solutions u to pde's of the form

$$(1.1) \qquad Lu = u_t - \nabla \cdot (A\nabla u) - B\nabla u = 0$$

under the following structure assumptions on A, B.

$$(1.2) \qquad \langle\, A(X,t)\xi, \xi\,\rangle \;\geq\; \gamma_1 |\xi|^2$$

for some $\gamma_1 > 0$, almost every $(X,t) \in U$ and all $n \times 1$ matrices ξ.

$$(1.3) \qquad \left(\sum_{i=0}^{n-1} x_0 \,|B_i| \;+\; \sum_{i,j=0}^{n-1} |A_{ij}| \right)(X,t) < M < \infty$$

for almost every $(X,t) \in U$. For some large $\rho > 0$,

$$(1.4) \qquad A \equiv \text{ constant matrix in } U \setminus Q_\rho(0,0)\,.$$

If

$$d\mu_1(X,t) = x_0 \,|B|^2(X,t)\, dX dt,$$

then μ_1 is a Carleson measure on U with

$$(1.5) \qquad \|\mu_1\| \leq \beta_1 < \infty.$$

Also

$$d\mu_2(X,t) = (x_0 \,|\nabla A|^2 + x_0^3 \,|A_t|^2)(X,t)\, dX dt,$$

is a Carleson measure on U with

$$(1.6) \qquad \|\mu_2\| \;\leq\; \beta_2 \;<\; \infty\,.$$

Moreover

$$(1.7) \qquad d\mu_2 / dX dt \;\leq\; \Lambda < \infty$$

for a.e $(X,t) \in U$. Next we assumed whenever $0 \leq i, j \leq n - 1$,

$$\frac{\partial A_{ij}}{\partial x_0} \;=\; \sum_{l=0}^{n-1} \langle\, e_l^{ij}\,,\, \tfrac{\partial}{\partial x_l} f_l^{ij}\,\rangle \;+\; g^{ij}$$

in the distributional sense where

$$(1.8) \qquad [\sum_{l=0}^{n-1} |e_l^{ij}| + |f_l^{ij}|]\,(X,t) \leq \Lambda < \infty$$

for a.e $(X,t) \in U$ and that

$$d\mu_3(X,t) = [\sum_{i,j=0}^{n-1} (\sum_{l=0}^{n-1} x_0\,|\nabla e_l^{ij}|^2 + x_0^{-1}\,|f_l^{ij}|^2) + |g^{ij}|]\,(X,t)\,dX\,dt$$

is a Carleson measure on U with

$$(1.9) \qquad \|\mu_3\| \leq \beta_3 < \infty.$$

In Theorem 2.13 of chapter I we showed that if the Carleson norms of the μ's are small enough and A is near enough a constant matrix, then the Radon-Nikodym derivative of parabolic measure with respect to a given point is in a certain L^2 reverse Hölder class. In this chapter we remove these restrictions but at the expense of a further smoothness assumption on A, B. To this end let

$$d\hat{\mu}_1(X,t) = \text{ ess sup } \{x_0\,|B|^2(Y,s) : (Y,s) \in Q_{x_0/2}(X,t)\}\,dX\,dt,$$

$$d\hat{\mu}_2(X,t) = \text{ ess sup } \{ x_0\,|\nabla A|^2(Y,s) + x_0^3\,|A_s|^2(Y,s) : (Y,s) \in Q_{x_0/2}(X,t)\}\,dX\,dt,$$

and set

$$d\hat{\mu}_3(X,t) = \text{ ess sup } \quad \{[\sum_{i,j=0}^{n-1} (\sum_{l=0}^{n-1} [(x_0\,|\nabla e_l^{ij}|^2 + x_0^{-1}\,|f_l^{ij}|^2) + |g^{ij}|]\,(Y,s)$$

$$: (Y,s) \in Q_{x_0/2}(X,t)\}\,dX\,dt.$$

With this notation we prove in chapter II,

Theorem 1.10 *Let A, B satisfy (1.2)-(1.9) and either (*) or both (**), (***).*

(*) $\qquad$ *(1.5)-(1.9) hold with μ_i replaced by $\hat{\mu}_i$, for $1 \leq i \leq 3$.*

(**) $\qquad$ *A has distributional second partials and B has distributional first partials which at $(X,t) \in U$ satisfy*

$$\sum_{i,j=0}^{n-1} x_0^2\,(|\frac{\partial^2 A}{\partial x_i\,\partial x_j}| + \sum_{i=0}^{n-1} (x_0^3\,|\frac{\partial^2 A}{\partial x_i \partial t}| + x_0^2\,|\frac{\partial B}{\partial x_i}|) + x_0^4|\frac{\partial^2 A}{\partial t^2}| + x_0^3\,|\frac{\partial B}{\partial t}|) < \Lambda_1 < \infty.$$

(***) $\qquad$ *f_l^{ij}, g^{ij} have distributional first partial derivatives and e_l^{ij} has distributional second partial derivatives for $0 \leq i,j \leq n-1$ which satisfy at $(X,t) \in U$*

$$\sum_{i,j=0}^{n-1} (\sum_{l=0}^{n-1} x_0^2\,|\frac{\partial}{\partial t} f_l^{ij}| + x_0\,|\nabla f_l^{ij}|) + x_0\,|\nabla g^{ij}| + x_0^2\,|\frac{\partial}{\partial t} g^{ij}|$$

$$+ \sum_{i,j=0}^{n-1} \sum_{k,m,l=0}^{n-1} x_0^2\,|\frac{\partial^2 e_l^{ij}}{\partial x_m\,\partial x_k}| + \sum_{i,j=0}^{n-1} \sum_{l,m=0}^{n-1} x_0^3\,|\frac{\partial^2 e_l^{ij}}{\partial x_m \partial t}| + \sum_{i,j=0}^{n-1} \sum_{l=0}^{n-1} x_0^4\,|\frac{\partial^2 e_l^{ij}}{\partial t^2}| < \Lambda_1.$$

Then the continuous Dirichlet problem for the pde in (1.1) always has a unique solution. Moreover, if ω denotes parabolic measure corresponding to (1.1), A, B, then $\omega(d, x, t + 2d^2, \cdot)$ is mutually absolutely continuous with respect to Lebesgue *measure on $Q_d(x,t)$. Also for some $p, 1 < p < \infty$, $\frac{d\omega}{dyds}(d, x, t+2d^2, \cdot) \in \alpha_p(Q_d(x,t))$ with*

$$\|\tfrac{d\omega}{dyds}(d, x, t + 2d^2, \cdot)\|_{\alpha_p(Q_d(x,t))} < c^{**} < \infty,$$

for all $(x, t) \in R^n$, $d > 0$. Here c^{**} depends on $\beta_1, \beta_2, \beta_3, \gamma_1, M, \Lambda, n$ and also possibly Λ_1.

Remark. As usual all partial derivatives of the various vector functions in Theorem 1.10 are taken componentwise. To outline the proof of this theorem suppose that (∗) of Theorem 1.10 is valid and parabolic measure ω corresponding to (1.1), A, B, exists. In Lemma 4.1 we show that if $\xi = \sum_{i=1}^{3} \hat{\mu}_i[(0, d) \times Q_d(x, t)]$ is small enough, say $\xi \leq \epsilon$, and if E Borel $\subset Q_d(x, t)$ is a large enough fraction of $Q_d(x, t)$ (depending on ϵ), then

(1.11) $c\,\omega(d, x, t + 2d^2, E) \geq 1$

for some $c = c(\epsilon, \gamma_1, M, \Lambda, n)$. To prove (1.11) one first observes from a maximum principle that it suffices to prove (1.11) with ω replaced by $\tilde{\omega}$ and E by E'. Here E' is a closed set with $E' \subset E$ and $\tilde{\omega}$ is parabolic measure for a certain parabolic sawtooth, $\Omega \subset U$ with $E' \subset \partial\Omega$. Now we shall choose Ω in such a way that the Carleson measures in Theorem 1.10 are ' small ' on $\Omega \cap [(0, d/2) \times Q_d(x, t)]$. Next we extend A, B restricted to the above intersection to functions A_1, B_1 on U, where A_1, B_1 satisfy the hypotheses of Theorem 2.13 of chapter I. Finally we estimate $\tilde{\omega}$ using Theorem 2.13 and the parabolic measure corresponding to A_1, B_1.

In Lemma 2.1 we show that (1.11) is valid without any smallness assumption on ξ, i.e when $0 < \epsilon < \infty$. In this case we shall use an induction type argument to reduce back to the case of small ξ considered in Lemma 2.1. Again two important ingredients in the reduction are comparison lemmas for parabolic measures (Lemmas 3.22 and 3.33), as well as our ability to extend A, B from certain parabolic sawtooths in such a way that the resulting extensions satisfy the hypotheses of Theorem 2.13. The comparison lemmas mentioned above do not follow readily from the work of [DJK] because in our lemmas one of the measures need not be doubling.

To show that (1.11) for $0 < \xi < \infty$ implies Theorem 1.10 we note that if $\hat{\mu}_i$ are Carleson measures, $1 \leq i \leq 3$, then from from the above discusssion, (1.11) holds whenever $d > 0, (x, t) \in R^n$ with constants independent of $Q_d(x, t)$. Second we observe that if we knew $\omega(d, x, t + 2d^2, \cdot)$ was a doubling measure, then it would follow from the above remark that Theorem 1.10 is true (see [CF]). Unfortunately we have been unable to prove that ω is doubling, primarily because the proof seems to rely on proving some basic estimates near ∂U similar to those in section 3 of chapter I for certain solutions to the adjoint pde corresponding to (1.1). We have in fact been unable to obtain any meaningful boundary estimates for the adjoint pde. As a consequence we are unable to use the method in [FS] to get parabolic doubling when the Carleson norms in (1.5), (1.6), (1.9) are large. To overcome this possible lack of doubling we show in Lemma 3.6 that (1.11) implies the conclusion of Theorem 1.10. The proof of Theorem 1.10 is given in sections 2, 3, and 4. For more discussion concerning doubling and also possible generalizations of this theorem we refer the reader to the remark at the end of section 4. In section 2 we also point out that one corollary of Theorem 1.10 is the following theorem of [LM, ch

3] mentioned in section 1 of chapter I.

Corollary 1.12 *Let ψ be as in (1.3), (1.4) of chapter I with compact support in R^n and put $\Omega = \{(x_0 + \psi(x,t), x, t) : (X,t) \in U\}$. Let ω be parabolic measure corresponding to the heat equation in Ω and let ρ be as in (1.6) of chapter I. Put $\hat{\omega}(X,t,E) = \omega(\rho(X,t), \rho(E))$ whenever $(X,t) \in U$ and $E \subset R^n$ is a Borel set. There exists $p, 1 < p < \infty$, such that $\frac{d\hat{\omega}}{dyds}(d, x, t + 2d^2, \cdot) \in \beta_p(Q_d(x,t))$ with*

$$\|\tfrac{d\hat{\omega}}{dyds}(d, x, t + 2d^2, \cdot)\|_{\beta_p(Q_d(x,t))} < c^+ < \infty,$$

for all $(x,t) \in R^n$, $d > 0$, where c^+ depends only on a_1, a_2 and n.

Remark. 1) Corollary 1.12 is stated in terms of Muckenhoupt weights in [LM, ch 3] and for ω rather than $\hat{\omega}$. However both statements are easily seen to be equivalent. To prove Corollary 1.12 we essentially need only show that a solution to the heat equation composed with the mapping ρ defined in (1.6) of chapter I is a solution in U to a pde satisfying the conditions of Theorem 1.10.
2) The proof of [LM] for absolute continuity of parabolic measure corresponding to the heat equation in a time-varying domain relies heavily on L^p estimates for some complicated singular integral operators. We shall completely avoid using singular integral theory.
3) As another application of Theorem 1.10 we show in section 4 that a certain $L^q(R^n)$ Dirichlet problem has a solution. More specifically, we prove

Theorem 1.13 *Let A, B, p be as in Theorem 1.10 and put $q_0 = p/(p-1)$. If $q_0 \leq q < \infty$ and $f \in L^q(R^n)$ with compact support, then there exists u a weak solution to (1.1) with*

$$(I) \qquad \lim_{(Y,s) \to (x,t)} u(Y,s) = f(x,t)$$

for almost every $(x,t) \in R^n$ where the limit is taken through $(Y,s) \in \Gamma(x,t)$. Also $Nu \in L^q(R^n)$ and

$$(II) \qquad\qquad \|Nu\|_{L^q(R^n)} \leq \hat{c}\|f\|_{L^q(R^n)}$$

where $\hat{c}$ has the same dependence as c^{**} in Theorem 1.10. u is the unique weak solution to (1.1) with properties (I) and (II).

In Theorem 1.13, N is the nontangential maximal function defined as in (2.14) of chapter I. In the elliptic case we can prove stronger versions of Theorems 1.10, 1.13. To do so define $\hat{\mu}_i, i = 1, 2$ as above Theorem 1.10 with $(X,t), dXdt$ replaced by X, dX (so $A_s \equiv 0$).

   In chapter III, section 4, we shall outline the proof of the following two theorems.

Theorem 1.14 *Let $A = A(X), B = B(X)$ satisfy (1.2) - (1.7) and either (1.5) - (1.7) with μ_i replaced by $\hat{\mu}_i, i = 1, 2$ or (**) of Theorem 1.10 with (X,t) replaced by X in $\hat{U} = \{X : x_0 > 0\}$. Then the continuous Dirichlet problem for the pde*

$$(+) \qquad \nabla \cdot (A\nabla u) + B\nabla u = 0$$

in $\hat{U}$ always has a unique weak solution. If ω denotes elliptic measure corresponding to (+), then $\omega(d, x, \cdot)$ is mutually absolutely continuous with respect to Lebesgue *measure on $B_d(x) = \{y : |y - x| < d\}$. Also for some $p, 1 < p < \infty$, $\frac{d\omega}{dy}(d, x, \cdot) \in \beta_p^*(B_d(x))$ with*

$$\|\tfrac{d\omega}{dy}(d, x, \cdot)\|_{\beta_p^*(B_d(x))} < c^{++} < \infty,$$

for all $x \in \mathbf{R}^{n-1}$, $d > 0$. Here c^{++} depends on the constants in (1.2)-(1.7) and possibly also (**) of Theorem 1.10. Moreover $\beta_p^*(B_d(x))$ stands for a strong reverse Hölder class defined in the same way as $\beta_p(Q_d(x, t))$ with $Q_d(x, t)$ replaced by $B_d(x)$.

Theorem 1.15.   *Let A, B, p be as in Theorem 1.14 and put $q_0 = p/(p-1)$. If $q_0 \le q < \infty$ and $f \in L^q(\mathbf{R}^{n-1})$ with compact support, then there exists u a weak solution to (+) in $\hat{U}$ with*

$$(I) \quad \lim_{Y \to x} u(Y) = f(x)$$

for almost every $x \in \mathbf{R}^{n-1}$ where the limit is taken through $Y \in \tilde{\Gamma}(x)$. Also $Nu \in L^q(\mathbf{R}^n)$ and

$$(II) \qquad \qquad \|Nu\|_{L^q(\mathbf{R}^n)} \le \bar{c}\|f\|_{L^q(\mathbf{R}^n)}$$

where $\bar{c}$ has the same dependence as c^{++} in Theorem 1.14. u is the unique weak solution to (+) with properties (I) and (II).

Remark. 1) We note that Theorem 1.14 has already been proved by Kenig and Pipher (oral communcation of Kenig) using a different method than ours. In fact an earlier version of this theorem required the integration by parts hypothesis (1.9) and (***). The new idea which allows us to do away with the integration by parts hypothesis (essentially that it suffices to consider A lower triangular) is garnered from reading [KKPT].
2) The cone $\tilde{\Gamma}(x)$ is defined similar to $\Gamma(x, t)$ in Theorem 1.13 and Nu is the nontangential maximal function of u relative to $\tilde{\Gamma}(x)$.
3) In chapter III, section 4, we shall show (see Lemma 4.6) that ω as above is a doubling measure. Thus in the elliptic case we can show $\frac{d\omega}{dyds}(d, x, \cdot)$ is in a strong reverse Hölder class.

2. Preliminary Reductions for Theorem 1.10

To begin the the proof of Theorem 1.10 for given $Q_d(x, t)$ let $\mu_i, 1 \le i \le 3$, be defined as in (1.5)-(1.9) in $(0, d) \times Q_d(x, t)$ and set

$$d\mu^*(Z, \tau) = \sum_{i=1}^{3} d\mu_i(Z, \tau) = L(Z, \tau) \, dZ d\tau$$

when $(Z, \tau) \in (0, d) \times Q_d(x, t)$, where

$$L(Z, \tau) = [\, z_0 \, |B|^2 + z_0 \, |\nabla A|^2 + z_0^3 \, |\tfrac{\partial A}{\partial \tau}|^2$$

$$+ \sum_{i,j=0}^{n-1} (\sum_{l=0}^{n-1} z_0 \, |\nabla e_l^{ij}|^2 + z_0^{-1} \, |f_l^{ij}|^2) + |g^{ij}| \,] (Z, \tau).$$

Also set

$$K(Z,\tau) = \text{ ess sup } \{L(Y,s) : (Y,s) \in Q_{z_0/2}(Z,\tau) \cap [(0,d) \times Q_d(x,t)]\,\}$$

when $(Z,\tau) \in (0,d) \times Q_d(x,t)$ and put

$$d\mu'(Z,\tau) = K(Z,\tau)dZd\tau$$

on $(0,d) \times Q_d(x,t)$. With this notation we prove a key lemma.

Lemma 2.1. *Let A, B satisfy (1.2)-(1.4) in U and suppose for some $(x,t) \in R^n, d > 0, \epsilon_2 > 0$ small that*

(i) $\dfrac{\partial}{\partial x_0} A_{ij} = \displaystyle\sum_{l=0}^{n-1} \langle e_l^{ij}, \dfrac{\partial}{\partial x_l} f_l^{ij} \rangle \, + \, g^{ij}$ *on $(0,\infty) \times Q_d(x,t)$ in the*

   distributional sense where e_l^{ij}, f_l^{ij} are vector functions with distributional partial derivatives on $(0,\infty) \times Q_d(x,t)$.

(ii)   *(1.7), (1.8) are valid at points in $(0,\infty) \times Q_d(x,t)$.*

(iii) $\tilde\mu[(0,d) \times Q_d(x,t)] \leq \epsilon_2 |Q_d(x,t)|$ *where either (a) $\tilde\mu = \mu'$ or (b) $\tilde\mu = \mu^*$ and (**), (***) of Theorem 1.10 hold with U replaced by $(0,d) \times Q_d(x,t)$.*

If $\epsilon_2, 0 < \epsilon_2 < \min\{\epsilon_1,\epsilon_0\}$, is small enough (depending only on γ_1, M, n, Λ and possibly Λ_1), there exists $\eta_0 = \eta_0(\epsilon_2), \eta_1 = \eta_1(\epsilon_2), 0 < \eta_0, \eta_1 < 1/2$, such that the following statement is true. Let $u, 0 \leq u \leq 2$, be a solution to (1.1) in U, corresponding to A, B as above, which is continuous on $\bar U$. If $u \equiv 1$ on some closed set $E \subset Q_d(x,t)$ with

$$|E| \geq (1 - \eta_0) |Q_d(x,t)|,$$

then

$$u(d, x, t + 2d^2) \geq \eta_1.$$

Proof: We emphasize that no Carleson measure assumptions are made on $\tilde\mu$ in Lemma 2.1. We write $d\tilde\mu(Z,\tau) = H(Z,\tau)\,dZd\tau$ for $(Z,\tau) \in (0,d) \times Q_d(x,t)$ and note that $H \equiv K$ when (a) holds while $H \equiv L$ when (b) of Lemma 2.1 is valid. Next let F be closed, $F \subset Q_d(x,t)$ and set

$$\delta = \epsilon_2^{1/[1000(n+2)]},$$

$$\hat\sigma(z,\tau,F) = \inf\{|z - y| + |s - \tau|^{1/2} : (y,s) \in F\},$$

$$\hat\Omega = \{(Z,\tau) \in U : z_0 > \delta^4\,\hat\sigma(z,\tau,F)\,\},$$

$$A_0 = \frac{4}{d\,|Q_d(x,t)|} \int_{(d/4,d/2) \times Q_d(x,t)} A(Z,\tau)\,dZd\tau,$$

where the integral is taken componentwise. We shall show for $\epsilon_2 > 0$ sufficiently small that there exists F as above with

(2.2) $|Q_d(x,t) \setminus F| \leq \delta\,|Q_d(x,t)|$

and

$$(+) \qquad \int_{\delta^4 \hat{\sigma}(z,\tau,F)}^{d/2} L(z_0, z, \tau) \, dz_0 \leq \delta^{100} \text{ for a.e } (z, \tau) \in Q_d(x, t),$$

(2.3)

$$(++) \qquad z_0 \, L(Z, \tau) \leq \delta^{40}, \text{ for } (Z, \tau) \in \hat{\Omega} \cap [(0, d/2) \times Q_d(x, t)],$$

$$(+++) \qquad |A - A_0|(Z, \tau) \leq \delta, \text{ for } (Z, \tau) \in \hat{\Omega} \cap [(0, d/2) \times Q_d(x, t)].$$

To prove (2.2), (2.3), we put $\epsilon = \epsilon_2$ and temporarily allow ϵ to vary. Put

$$k(z, \tau) = \int_0^d H(z_0, z, \tau) \, dz_0, \ (z, \tau) \in Q_d(x, t),$$

and set $k \equiv 0$ otherwise in $R^n \setminus Q_d(x, t)$. Note from the assumptions in Lemma 2.1 that

$$\int_{Q_d(x,t)} k \, dz d\tau \leq \epsilon |Q_d(x, t)|.$$

From this remark and weak type estimates we see there exists $F_1 \subset Q_d(x, t)$, with F_1 closed and

(2.4)

$$|Q_d(x, t) \setminus F_1| \leq \epsilon^{1/2} |Q_d(x, t)|$$

$$Mk(z, \tau) \leq c(n) \epsilon^{1/2}, \ (z, \tau) \in F_1,$$

where as usual $Mk(z, \tau)$ is the Hardy - Littlewood maximal function of k taken with respect to rectangles containing (z, τ) of length ρ, ρ^2 in the space and time variables, respectively. Again from weak type estimates we see for ϵ small enough that there exists $F_2 \subset F_1$, F_2 closed such that

(2.5)

$$|Q_d(x, t) \setminus F_2| \leq \epsilon^{1/4} |Q_d(x, t)|,$$

$$M(\chi)(z, \tau) \leq c(n) \epsilon^{1/4}, \ (z, \tau) \in F_2,$$

where χ denotes the characteristic function of $Q_d(x, t) \setminus F_1$.

We assume as we may, thanks to (1.7), that A is locally Lipschitz continuous in U. We claim that if $T = (\delta^4 d, d/2) \times Q_d(x, t)$, then

(2.6)

$$\sup_{(Z,\tau) \in T} |A - A_0|(Z, \tau) \leq \delta^2$$

for $0 < \delta < \delta_0$, where δ_0 depends only on γ_1, M, Λ, and possibly Λ_1. To prove this claim observe from (2.4) that if (a) of Lemma 2.1 holds, then for $(Z, \tau) \in T$ with $(z, \tau) \in F_1$, we have

$$\text{ess sup } \{ y_0 \, |\nabla A|^2(Y, s) + y_0^3 \, |\tfrac{\partial}{\partial t} A|^2(Y, s) : (Y, s) \in Q_{z_0/16}(Z, \tau) \}$$

(2.7)

$$\leq 2 z_0^{-1} \int_{z_0}^{3z_0/2} H(\hat{z}_0, z, \tau) \, d\hat{z}_0 \leq c(n) \epsilon^{1/2} / (\delta^4 d).$$

Since

$$|Q_d(x, t) \setminus F_1| \leq \epsilon^{1/2} 2^n d^{n+1} \leq c(n) (\delta^{500} d)^{n+1}$$

we deduce first for ϵ small enough that T is contained in the union of rectangles of the form $Q_{z_0/16}(Z,\tau)$, with $(Z,\tau) \in T$ and $(z,\tau) \in F_1$. Second we deduce from (2.7) that for almost every $(Z,\tau) \in T$, we have

$$z_0\,|\nabla A|^2(Z,\tau) \,+\, z_0^3\,|\tfrac{\partial}{\partial\tau}A|^2(Z,\tau) \,\leq\, c(n)\,\epsilon^{1/2}\,/(\delta^4\,d).$$

Now suppose that (b) of Lemma 2.1 is valid. Then we assume, as we may, thanks to (**), that $\nabla A, \tfrac{\partial}{\partial t}A$, are locally Lipschitz continuous in $(0,d) \times Q_d(x,t)$. From weak type estimates and (2.4) we find for fixed $(\hat{z},\hat{\tau}) \in F_1$, and $0 < \hat{z}_0 < d/2$, that

$$(2.8) \qquad \hat{z}_0\,|\nabla A|^2(\hat{Z},\hat{\tau}) \,+\, \hat{z}_0^3\,|\tfrac{\partial}{\partial\tau}A|^2(\hat{Z},\hat{\tau}) \,\leq\, c(n)\,\epsilon^{1/4}\,/d,$$

except for a set of $\hat{z}_0 \in (0,d/2)$ of measure at most $\epsilon^{1/4}\,d$. From (2.4) we get that the Lebesgue $n+1$ measure of the set $\subset T$ where (2.8) does not hold is at most $c\,\epsilon^{1/4}\,d^{n+2}$. Thus given $(Z,\tau) \in T$ there exists $(\hat{Z},\hat{\tau})$ for which (2.8) holds and with

$$(2.9) \qquad |\hat{Z} - Z| + |\hat{\tau} - \tau|^{1/2} \,\leq\, c\,\epsilon^{\frac{1}{4(n+2)}}\,d.$$

Using (2.8), (2.9), and smoothness assumption (**) on the second derivatives of A, we find that at $(Z,\tau) \in T$

$$(2.10) \qquad z_0\,|\nabla A|^2 \,+\, z_0^3\,|\tfrac{\partial}{\partial\tau}A|^2 \,\leq\, c\,(\epsilon^{1/4}/d \,+\, \epsilon^{\frac{1}{4(n+2)}}\,d/z_0^2) \,\leq\, \delta^{200}\,/d$$

for $\epsilon = \epsilon(M,\Lambda,\Lambda_1,n) > 0$ small enough. In either case we conclude that (2.10) is valid when $(Z,\tau) \in T$. Using (2.10) and basic Sobolev estimates we find that claim (2.6) is true.

Next we show for given $\rho > 0$ that the function $(z,\tau) \to A(\rho,z,\tau)$ converges as $\rho \to 0$ in $L^2(Q_d(x,t))$ to a function denoted $A(0,\cdot)$ with

$$(2.11) \qquad \int_{Q_d(x,t)} (A(0,z,\tau) - A_0)^2\,dz\,d\tau \,\leq\, c\,\delta^4\,|Q_d(x,t)|,$$

where $c = c(M,\Lambda,\Lambda_1,n)$. To prove this inequality let $0 < \rho_1 < \rho_2 \leq d/2$, and $Q_r(y,s) \subset Q_d(x,t)$. Now for almost every ρ_1,ρ_2,r with respect to one dimensional Lebesgue measure and pairs (i,j), with $0 \leq i,j \leq n-1$, we see from assumption

(i) of Lemma 2.1 that

(2.12)

$$I = \int_{Q_r(y,s)} (A_{ij}(\rho_2, z, \tau) - A_{ij}(\rho_1, z, \tau))^2 \, dz \, d\tau$$

$$= 2 \int_{\rho_1}^{\rho_2} \int_{Q_r(y,s)} (A_{ij}(\rho, z, \tau) - A_{ij}(\rho_1, z, \tau)) \tfrac{\partial}{\partial \rho} A_{ij}(\rho, z, \tau) \, dz d\tau d\rho$$

$$= 2 \int_{\rho_1}^{\rho_2} \int_{Q_r(y,s)} (A_{ij}(\rho, \cdot) - A_{ij}(\rho_1, \cdot)) [(\sum_{l=0}^{n-1} \langle e_l^{ij}, \tfrac{\partial}{\partial z_l} f_l^{ij} \rangle) + g^{ij}](\rho, \cdot) \, dz d\tau d\rho$$

$$= 2 \int_{\rho_1}^{\rho_2} \int_{Q_r(y,s)} (A_{ij}(\rho, \cdot) - A_{ij}(\rho_1, \cdot)) [-(\sum_{l=0}^{n-1} \langle \tfrac{\partial}{\partial z_l} e_l^{ij}, f_l^{ij} \rangle) + g^{ij}](\rho, \cdot) dz d\tau d\rho$$

$$- 2 \int_{\rho_1}^{\rho_2} \int_{Q_r(y,s)} \sum_{l=0}^{n-1} \tfrac{\partial}{\partial z_l} [A_{ij}(\rho, \cdot) - A_{ij}(\rho_1, \cdot)] \langle e_l^{ij}, f_l^{ij} \rangle (\rho, \cdot) \, dz d\tau d\rho$$

$$+ \int_{\partial[(\rho_1,\rho_2) \times Q_r(y,s)]} (A_{ij}(\rho, \cdot) - A_{ij}(\rho_1, \cdot)) [\sum_{l=0}^{n-1} \langle e_l^{ij}, f_l^{ij} \rangle \langle e_l, \nu \rangle](\rho, \cdot) d\xi$$

$$= S_1 + S_2 + S_3.$$

In the last integral ξ denotes surface area, ν is the outer unit normal to $(\rho_1, \rho_2) \times Q_r(y,s)$, and e_l is a unit vector parallel to the x_l axis. Using Cauchy's inequality, (1.3), (1.8) and (2.4) we deduce that if $Q_r(y,s) \cap F_1 \neq \emptyset$, then

(2.13) $\qquad |S_1| + |S_2| \le c \int_{[\rho_1,\rho_2] \times Q_r(y,s)} |H|(Z,\tau) dZ \, d\tau \le c \, \epsilon^{1/2} |Q_r(y,s)|.$

From (2.13) and integrability of H we get

(2.14) $\qquad\qquad\qquad |S_1| + |S_2| \to 0$ as $\rho_1, \rho_2 \to 0$

outside a set of linear measure zero. Also, from (1.8) we have

(2.15) $\qquad |S_3| \le c r^n (\rho_2 - \rho_1) + c \sum_{k=1}^{2} \int_{Q_r(y,s)} |f_0^{ij}| (\rho_k, z, \tau) \, dz \, d\tau.$

If $f(\rho) = \int_{Q_r(y,s)} |f_0^{ij}| (\rho, z, \tau) \, dz \, d\tau$, we shall show that

(2.16)
$\qquad (\alpha) \qquad f(\rho) \to 0$ as $\rho \to 0$ outside a set of linear measure zero ,

$\qquad (\beta) \qquad f(\rho) \le c\epsilon^{1/6} r^{n+1},$

for some $c = c(M, \Lambda, \Lambda_1, n)$ provided $F_1 \cap Q_r(y,s) \neq \emptyset$ and $0 < \rho < d/2$. To prove (2.16) first assume that (a) of Lemma 2.1 holds. Then from Cauchy's inequality we see for almost every $\rho \in (0, d/2)$ that

$$f(\rho)^2 \le c r^{n+1} \int_{\rho}^{3\rho/2} \int_{Q_r(y,s)} H(z_0, z, \tau) dz_0 \, dz \, d\tau = I_1.$$

Now, $I_1 \leq c\epsilon^{1/2} r^{2n+2}$ independently of ρ thanks to (2.4). Thus (2.16) holds in this case. If (b) holds, then from $(***)$ of Theorem 1.10 we see for small $\eta > 0$ that $|f(\rho) - f(\rho')| \leq c\eta r^{n+1}$ when $|\rho - \rho'| \leq \eta\rho$. Using this fact and arguing as above, we get

$$f(\rho)^2 \leq c\eta^2 r^{2n+2} + c\eta^{-1} r^{n+1} \int_\rho^{(1+\eta)\rho} \int_{Q_r(y,s)} H(z_0, z, \tau) dz_0 dz d\tau.$$

Since H is integrable we conclude first that $\limsup_{\rho \to 0} f(\rho)^2 \leq c r^{2n+2} \eta^2$ and second from arbitrariness of η that the limit in (2.16) exists. Finally choosing $\eta = \epsilon^{1/6}$ in the above inequality and using (2.4) we get (2.16).

We note that I in (2.12) is continuous as a function of ρ_1, ρ_2 as follows from (1.7). Using this remark, (2.14), (2.15), and (2.16)(α) with $r = d$, we conclude that $A(\rho, \cdot) \to A(0, \cdot)$ in the norm of $L^2(Q_d(x,t))$. Moreover, letting $\rho_2 = \delta^4 d$, $r = d$ and $\rho_1 \to 0$ in (2.12) we deduce from (2.13), (2.15), (2.16)(β), (2.6) that (2.11) holds for δ sufficiently small. From (2.11) and the Hardy - Littlewood maximal theorem we get the existence of $F \subset F_2$, F closed with

$$|Q_d(x,t) \setminus F| \leq \delta |Q_d(x,t)|$$

(2.17)

$$M[(A(0, \cdot) - A_0)^2] \leq c\delta^3, \quad (z, \tau) \in F,$$

where c has the same dependence as the constant in (2.11). Clearly (2.2) holds for F. Next we prove (2.3) $(+)$. Let $(z, \tau) \in Q_d(x,t)$ and put $r = \hat{\sigma}(z, \tau, F)$. Then from (2.5) we see there exists $(z_1, \tau_1) \in F_1$ with $|z - z_1| + |\tau_1 - \tau|^{1/2} \leq \delta^{200} r$ for ϵ small enough. If (a) of Lemma 2.1 is valid, then from (2.4) we find for almost every $(z, \tau) \in Q_d(x,t)$ with respect to Lebesgue n measure that

$$\int_{\delta^4 r}^{d/2} L(z_0, z, \tau) \, dz_0 \leq \int_{\delta^4 r}^{d/2} H(z_0, z_1, \tau_1) \, dz_0 \leq c\epsilon^{1/2}.$$

If (b) of Lemma 2.1 holds, we use $(**), (***)$ to get that

$$\int_{\delta^4 r}^{d/2} L(z_0, z, \tau) \, dz_0 \leq \int_{\delta^4 r}^{d/2} L(z_0, z_1, \tau_1) \, dz_0$$

$$+ c\delta^{200} r \int_{\delta^4 r}^{d/2} z_0^{-2} \, dz_0 \leq \delta^{100}$$

for $\delta > 0$ sufficiently small. Thus in either case (2.3) $(+)$ is valid. To prove (2.3) $(++)$ for L we argue as in (2.7)-(2.10). Indeed, in case (a), one gets (2.3) $(++)$ by essentially repeating the above argument. In case (b) we use (2.3) $(+)$, weak type estimates, and $(**), (***)$ of Theorem 1.10 to get (2.3) $(++)$. To prove (2.3) $(+++)$ let (z_1, τ_1) in $Q_d(x,t) \cap F$. Then using (2.17), (2.12)-(2.16) with $(y, s) = (z_1, \tau_1)$, $\rho_2 = 2n\delta^4 r$, $\rho_1 = 0$, we find for $0 < r \leq d$ that

$$\int_{Q_r(z_1, \tau_1) \cap Q_d(x,t)} [A(2n\delta^4 r, y, s) - A_0]^2 \, dy ds \leq c\delta^3 r^{n+1}.$$

Since $\{2n\delta^4 r\} \times Q_r(z_1, \tau_1) \subset \hat{\Omega}$ we conclude from (2.3) $(++)$ that for some $c = c(M, \Lambda, \Lambda_1, n)$,

(2.18) $|A - A_0| \leq c\delta^{3/2}$ on $[\{2n\delta^4 r\} \times Q_r(z_1, \tau_1)] \cap [(0, d/2) \times Q_d(x,t)]$.

If $(Z,\tau) \in \hat{\Omega}$ and $z_0 \geq \delta^4 d$, then (2.6) implies (2.3)(+++). Otherwise we can choose $(z_1, \tau_1) \in F$ with

$$|z - z_1| + |\tau - \tau_1|^{1/2} = \hat{\sigma}(z, \tau, F)$$

and put $2n\delta^4 r = z_0$. Using (2.18) and possibly also (2.3) (++) we obtain (2.3)(+++) for δ sufficiently small.

Armed with (2.2), (2.3) we are now ready to use Theorem 2.13, Lemma 3.37, and Lemma 3.9 of chapter I to complete the proof of Lemma 2.1. Let $E \subset \bar{Q}_\rho(x,t)$ be closed and let $\hat{\sigma} = \hat{\sigma}(\cdot, E)$ be the parabolic distance function defined above (2.2). We claim there exists a regularized distance function $\sigma(\cdot, E) \in C_0^\infty(Q_{4d}(x,t))$ with the following properties:

(a) $c^{-1}\hat{\sigma} \leq \sigma \leq c\,\hat{\sigma}, \quad \text{on } Q_{2d}(x,t),$

(b) $|\sigma(x,t) - \sigma(y,s)| \leq c\,[\,|x-y| + |s-t|^{1/2}\,], \; (x,t),(y,s) \in R^n,$

(2.19)

(c) σ has distributional partial derivatives in x,t on $R^n \setminus E$ and
$\hat{\sigma}^{l-1}|\frac{\partial^l}{\partial x_k^l}\sigma|(x,t) + \hat{\sigma}^{2l-1}|\frac{\partial^l}{\partial t^l}\sigma|(x,t) \leq c.$ for $(x,t) \in R^n \setminus E,$
$0 \leq k \leq n-1$ and l a positive integer.

Proof: In (2.19), $c = c(l,n)$. The construction of a regularized distance function in the usual Euclidean case is more or less standard (see [St, ch 6]). The only difference in the parabolic case is that one uses a partition of unity adapted to a Whitney decomposition of $R^n \setminus E$ into rectangles (of side length ρ, ρ^2 in the space and time variables) rather than cubes. $\square$

Next let $\hat{\xi} \in C^\infty(R), 0 \leq \hat{\xi} \leq 1$, with supp $\hat{\xi} \subset (1/2, \infty), \hat{\xi} \equiv 1$ on $(1, \infty)$, and $|\hat{\xi}'| \leq 100$. For $(Z,\tau) \in U$ let

$$\xi(Z,\tau) \equiv \begin{cases} \hat{\xi}\left(\frac{z_0}{\delta^3\sigma(z,\tau,F)}\right) & \text{for } \sigma(z,\tau,F) \neq 0 \\[2mm] = 1 \text{ when } \sigma(z,\tau,F) = 0. \end{cases}$$

We note that $\xi \in C^\infty(U)$ and if $(Z,\tau) \in U, then$

$$z_0^l|\frac{\partial^l}{\partial z_k^l}\xi(Z,\tau)| + z_0^{2l}|\frac{\partial^l}{\partial\tau^l}\xi(Z,\tau)| \leq c(l,n),$$

(2.20)

$$\text{supp } [|\frac{\partial^l}{\partial z_k^l}\xi| + |\frac{\partial^l}{\partial\tau^l}\xi|] \subset \{(Z,\tau) \in U : \delta^3\,\sigma(z,\tau) \leq 2z_0 \leq 2\delta^3\sigma(z,\tau)\},$$

for l a positive integer and $0 \leq k \leq n-1$, thanks to (2.19)(c). Also let $\theta \in C_0^\infty[(-d/2, d/2) \times Q_{d/2}(x,t)], 0 \leq \theta \leq 1$, with $\theta \equiv 1$ on $(-d/4, d/4) \times Q_{d/4}(x,t)$ and

$$d\,\|\nabla\theta\|_{L^\infty(\mathbb{R}^{n+1})} + d^2\,\|\frac{\partial}{\partial t}\theta\|_{L^\infty(\mathbb{R}^{n+1})} \leq c(n)\,.$$

Next define A_1, B_1 on U by

$$A_1 = (A - A_0)\,\theta^2\,\xi^2 + A_0$$

$$B_1 = B\,\theta\,\xi$$

where A_0 is the constant matrix in (2.3). We claim that A_1, B_1 satisfy the hypotheses of Theorem 2.13 for $\delta = \delta(\gamma_1, M, \Lambda, \Lambda_1, n, \epsilon_1)$ small enough. This claim is easily verified using (2.19), (2.20), and (2.3). For completeness we prove the most difficult assertion, (1.9) for A_1. For fixed (i,j), $0 \leq i, j \leq n-1$, observe that at $(Z, \tau) \in (0, \infty) \times Q_d(x,t)$,

$$\frac{\partial}{\partial z_0}(A_1)_{ij} = (A - A_0)_{ij} \theta^2 \frac{\partial}{\partial z_0} \xi^2 + (A - A_0)_{ij} (\frac{\partial}{\partial z_0} \theta^2) \xi^2 + (\frac{\partial}{\partial z_0}(A - A_0)_{ij}) \theta^2 \xi^2$$

$$= (A - A_0)_{ij} \theta^2 \frac{\partial}{\partial z_0} \xi^2 + (A - A_0)_{ij} (\frac{\partial}{\partial z_0} \theta^2) \xi^2 + \theta^2 \xi^2 (g^{ij} + \sum_{l=0}^{n-1} \langle e_l^{ij} \frac{\partial}{\partial z_l} f_l^{ij} \rangle)$$

$$= \tilde{g}^{ij} + \sum_{i=0}^{n-1} \langle \tilde{e}_l^{ij} \frac{\partial}{\partial z_l} \tilde{f}_l^{ij} \rangle$$

where

$$\tilde{e}_l^{ij} = \delta^2 e_l^{ij} \theta \xi$$

$$\tilde{f}_l^{ij} = \delta^{-2} f_l^{ij} \theta \xi$$

$$\tilde{g}^{ij} = (A - A_0)_{ij} \theta^2 \frac{\partial}{\partial z_0} \xi^2 + (A - A_0)_{ij} (\frac{\partial}{\partial z_0} \theta^2) \xi^2$$

$$+ g^{ij} \theta^2 \xi^2 - \sum_{l=0}^{n-1} \langle e_l^{ij}, f_l^{ij} \rangle \frac{\partial}{\partial z_l}(\theta \xi).$$

Using (2.3), (2.19), and (2.20), it is easily checked that μ_3 as in (1.9) has norm $\leq c(n) \delta$. Thus A_1, B_1 satisfy the hypotheses of Theorem 2.13 for δ sufficiently small (depending on ϵ_0).

We now prove Lemma 2.1. Let $\eta_0 = \delta$ and note from (2.2) that if $E \subset Q_d(x,t)$ is closed with $|E| \geq (1 - \delta) |Q_d(x,t)|$, then $|E \cap F| \geq (1 - 2\delta) |Q_d(x,t)|$. Thus if χ_1 denotes the characteristic function of $Q_d(x,t) \setminus (E \cap F)$, and $G = \{(z, \tau) : M(\chi_1)(z, \tau) \leq \delta^{1/2}\}$, then for δ sufficiently small

$$|Q_d(x,t) \setminus G| \leq \tfrac{1}{2} |Q_{d/8}(x,t)|.$$

Let ω_1 be parabolic measure corresponding to A_1, B_1. From Theorem 2.13 we see that if $Q_r(y,s) \subset R^n$, then $\frac{d\omega_1}{dyds}(r, y, s + 2r^2, \cdot) \in \beta_2(Q_r(y,s))$ with reverse Hölder constant $\leq c^*(\epsilon_0, \gamma_1, M, \Lambda, n)$. Using this fact it is not difficult to show that $\omega_1(r, y, s + 2r^2, \cdot)$ is an A_∞ weight with respect to Lebesgue measure on $Q_r(y,s)$ where rectangles are used instead of cubes in the usual definition(see [CF]). Also from Lemma 3.37 of chapter I, we have $c \omega_1(r, y, s + 2r^2, Q_r(y,s)) \geq 1$. Next let $\tilde{\psi}(z, \tau) = \delta^2 \sigma(z, \tau, F)$ and set

$$\tilde{\rho}(Z, \tau) = (z_0 + P_{z_0} \tilde{\psi}(z, \tau), z, \tau), \ (Z, \tau) \in U,$$

where P is as in section 1 of chapter I. We note from (2.19) and the remark after Lemma A in section 1 of chapter I that if δ is small enough, then $\tilde{\rho}$ maps U one to one and onto a region $\tilde{\Omega}$ with

(2.21) $\qquad A_1 = A, \ B_1 = B, \ \text{in } \tilde{\Omega} \cap [(0, 3d/16) \times Q_{d/4}(x,t)].$

Let v be a weak solution to (1.1) corresponding to A_1, B_1. We assert for δ sufficiently small that $v \circ \tilde{\rho}$ is a weak solution to (1.1) corresponding to some A_2, B_2 where A_2, B_2 satisfy (1.2)-(1.4) and (3.13) of chapter I. Thus our assertion imples that there exists parabolic measure ω_2 corresponding to (1.1), A_2, B_2 and Lemma 3.37 of chapter I holds for ω_2. We reserve the proof of this assertion until after (2.28).

Let u_2 be the solution to the continuous Dirichlet problem for (1.1), A_2, B_2, with $u_2 \equiv u \circ \tilde{\rho}$ on ∂U. Existence of u_2 follows from the above assertion. Next we use this assertion to show that for some $\hat{c}_1 = \hat{c}_1(\gamma_1, M, \Lambda, \Lambda_1, n) \geq 1$ we have

$$(2.22) \qquad \hat{c}_1 \, u_2(r, y, s + 2r^2) \geq \hat{c}_1 \, \omega_2(r, y, s + 2r^2, Q_r(y, s) \cap E \cap F) \geq 1.$$

whenever $(y, s) \in Q_{d/8}(x, t) \cap G$ and $0 < r \leq d/16$. To prove (2.22) first choose $E' \subset E \cap F \cap Q_r(y, s)$, with E' closed and

$$(2.23) \qquad |Q_r(y, s) \setminus E'| \leq 2|Q_r(y, s) \setminus (E \cap F)| \leq 2\,\delta^{1/2}\,|Q_r(y, s)|,$$

where the last inequality follows from the definition of G. We divide $Q_r(y, s) \setminus E'$ into a sequence of closed rectangles $\{\bar{Q}_j\}$ with disjoint interiors and whose side lengths in the space direction are proportional to their parabolic distance from E'. Let

$$\omega^*(Z, \tau, K) = \omega_1(\tilde{\rho}(Z, \tau), K), \quad K = \text{ Borel set } \subset R^n.$$

Then From Lemma 3.37(α) of chapter I for ω_1 and Harnack's inequality we see that if $Q_j = Q_{\hat{r}}(\hat{y}, \hat{s})$, then $c\,\omega^*(\cdot, Q_j) \geq 1$ on $Q_{\hat{r}/2}(\hat{y}, \hat{s})$. Also $\omega^*(\cdot, Q_j)$ satisfies (1.1) relative to A_2, B_2. Using these facts, the definiton of ω_2, and Lemma 3.37(β) for ω_2 we find that for each j

$$c^*\,\omega^*(r, y, s + 2r^2, \bar{Q}_j) \geq \omega_2(r, y, s + 2r^2, \bar{Q}_j),$$

where c^* has the same dependence as $\hat{c}_1$. Now using this inequality, the A_∞ property of ω_1, (2.23), and Lemma 3.37 for ω_2 we get for $\delta > 0$ small enough that

$$\omega_2(r, y, s + 2r^2, \bigcup \bar{Q}_i) \leq c^*\,\omega^*(r, y, s + 2r^2, \bigcup \bar{Q}_i) \leq \tfrac{1}{2}\,\omega_2(r, y, s + 2r^2, Q_r(y, s)).$$

Thus

$$2\,\omega_2(r, y, s + 2r^2, E') \geq \omega_2(r, y, s + 2r^2, Q_r(y, s)) \geq c^{-1}$$

for some c having the same dependence as $\hat{c}_1$, thanks to Lemma 3.37(α). Clearly this inequality implies the righthand inequality in (2.22). The lefthand inequality in (2.22) follows from the definition of ω_2 and the fact that $u_2 \equiv 1$ on $E \cap F$.

Next we note from the chain rule and (2.21) that $u \circ \tilde{\rho}$, u_2 satisfy the same pde in $(0, 3d/16) \times Q_{d/4}(x, t)$ for δ sufficiently small (see (2.28) for computations). Also these two functions agree on ∂U by our construction. From this observation, the above assertion, and the remark after (3.22) of chapter I we see that Lemma 3.9 of chapter I can be applied for ϵ_2 sufficiently small with u replaced by $u_2 - u \circ \tilde{\rho}$. Using this lemma we deduce the existence of $\hat{c}_2 \geq 8$ with the same dependence as $\hat{c}_1$ such that if $r = d/\hat{c}_2$, and $(y, s) \in G \cap Q_{d/8}(x, t)$, then

$$(2.24) \qquad \hat{c}_1| \, u_2 - u \circ \tilde{\rho}|(r, y, s + 2r^2) \leq 1/2.$$

Using (2.22), (2.24), we find that $\hat{c}_1 \, u \circ \tilde{\rho}(r, y, s + 2r^2) \geq 1/2$. From this inequality and Harnack's inequality we conclude first that $c\,u(d, x, t + 2d^2) \geq 1$ for some positive c depending only on $\epsilon_2, \gamma_1, M, \Lambda, \Lambda_1$, and second that Lemma 2.1 is true once we prove the assertion following (2.21).

Proof of Corollary 1.12. We prove the above assertion and Corollary 1.12 together since both follow easily from studying how (1.1) is transformed under the Dahlberg-Kenig-Stein transformation mentioned in section 1. Let $\psi : R^n \to R$ have compact support and satisfy

$$(2.25) \qquad |\psi(x, t) - \psi(y, s)| \leq a_1\,(\,|x - y| + |s - t|^{1/2}\,),$$

for some a_1, $0 < a_1 < \infty$. As in section 1 we put

$$\rho(X,t) = (x_0 + P_{\gamma x_0}\psi(x,t),\, x,\, t)$$

when $(X,t) \in U$ and note from (2.25), Lemma A in section 1, that $\gamma = \gamma(a_1, n) > 0$ can be chosen so small that

(2.26)
$$|\tfrac{\partial}{\partial x_0} P_{\gamma x_0}\psi(X,t)| \leq 1/2$$

whenever $(X,t) \in U$. Fix $\gamma > 0$ to be the largest number so that (2.26) holds and set

$$\Omega = \{\rho(X,t) : (X,t) \in U\}$$

First from properties of parabolic approximate identities we see that ρ has a continuous extension to ∂U defined by

$$\rho(x,t) = (\psi(x,t), x, t),\ (x,t) \in R^n,$$

and second from (2.26) we see that ρ maps $\bar{U}$ one to one and onto $\bar{\Omega}$. Let $\tilde{h}$ be a local solution to (1.1) in Ω corresponding to $\tilde{A}, \tilde{B}$ (for terminology see the discussion following (3.2) of chapter I) where $\tilde{A}, \tilde{B}$ satisfy (1.2)-(1.4) in Ω. Then a straightforward calculation shows that $h = \tilde{h} \circ \rho$ is a weak solution in $\bar{U}$ to (1.1) corresponding to A, B. Here A, B are defined as follows. For $(X,t) \in U$ let

$$\lambda(X,t) = \left(1 + \tfrac{\partial}{\partial x_0} P_{\gamma x_0}\psi\right)^{-1}(x,t)$$

and let $C(X,t) = (c_{ij}(X,t))$, $(X,t) \in U$, be the n by n matrix function with entries at (X,t),

$$c_{00} = \lambda,$$

(2.27)
$$c_{i0} = -\lambda \tfrac{\partial}{\partial x_i} P_{\gamma x_0}\psi,\ 1 \leq i \leq n-1,$$

$$c_{ii} = 1,\ 1 \leq i \leq n-1,$$

$$c_{ij} = 0 \text{ when } i \neq j \text{ and } 1 \leq j \leq n-1,\ 0 \leq i \leq n-1.$$

Then from the chain rule we find first that at (X,t)

$$\nabla \tilde{h} \circ \rho = C\,\nabla h$$

and second that

$$A = C^\tau (\tilde{A} \circ \rho)\, C$$

(2.28)
$$B = (\tilde{B} \circ \rho)\, C + \lambda\, (\nabla \tfrac{\partial}{\partial x_0} P_{\gamma x_0}\psi)^\tau A + \lambda \tfrac{\partial}{\partial t} P_{\gamma x_0}\psi\, e_0.$$

Here C^τ denotes the transpose of C, $e_0 = (1 \ldots 0)$ is a 1 by n row matrix, and the gradient of the above function is an n by 1 column matrix. We now prove the assertion following (2.21). Let $\tilde{A} = A_1$, $\tilde{B} = B_1$. Replacing $\psi, \rho, \tilde{h}, h$ by $\tilde{\psi}, \tilde{\rho}, v, v \circ \tilde{\rho}$, we see that (2.28) holds with $A_2 = A, B_2 = B$. Using this fact, (2.19), the remark after Lemma A in section 1 of chapter I and the fact that A_1, B_1 satisfy the hypotheses of Theorem 2.13, it is easily checked that A_2, B_2 satisfy (1.2)-(1.4) and (3.13) of chapter I. The proof of our assertion and Lemma 2.1 are now complete. $\square$

To prove Corollary 1.12, let ψ satisfy in addition to (2.25) the condition that

(2.29)
$$\|D^t_{1/2}\psi\|_* \leq a_2.$$

Let ω be parabolic measure corresponding to the heat equation in Ω. Given $E \subset R^n$ a Borel set put

$$\hat{\omega}(X, t, E) = \omega(\rho(X, t), \rho(E)), \ (X, t) \in U .$$

Then $\hat{\omega}$ is parabolic measure corresponding to (1.1) for A, B as defined in (2.28) with $\tilde{A}$ equal to the n by n identity matrix and $\tilde{B}$ equal to the 1 by n zero matrix. We shall show that this A, B satisfy the hypotheses of Theorem 1.10. To do so we note from (2.25)-(2.29) and Lemma A in section 1 of chapter I that (1.2)-(1.4) are valid. Also we have

(2.30)

(a) \quad If $\alpha = (\alpha_0, \dots, \alpha_{n-1})$ is a multi index and k a nonnegative integer, then
$$\|x_0^{-|\alpha|-2k} \tfrac{\partial^{|\alpha|+k}}{\partial x^\alpha \partial t^k} c_{ij}\|_{L^\infty(U)} \le c(a_1, a_2, |\alpha|, k, n), < \infty, 0 \le i, j \le n-1,$$

(b) \quad $d\nu(X, t) = [\, x_0 \, |\nabla c_{ij}|^2 + x_0^3 | \tfrac{\partial}{\partial t} c_{ij} |^2 \,] \, (X, t) \, dX \, dt$ is a Carleson measure on U with $\|\nu\| \le c(a_1, a_2)$,

(c) \quad $\tfrac{\partial}{\partial x_0} c_{00} = -\lambda^2 \tfrac{\partial^2}{\partial x_0^2} P_{\gamma x_0} \psi,$

(d) \quad $\tfrac{\partial}{\partial x_0} c_{i0} = \lambda^2 (\tfrac{\partial^2}{\partial x_0^2} P_{\gamma x_0} \psi) \tfrac{\partial}{\partial x_i} P_{\gamma x_0} \psi - \lambda \tfrac{\partial^2}{\partial x_i x_0} P_{\gamma x_0} \psi, \ 1 \le i \le n-1,$

(e) \quad $\tfrac{\partial}{\partial x_0} c_{ij} \equiv 0$ for $1 \le j \le n-1, \ 0 \le i \le n-1.$

From $(2.30)(a), (b)$, Lemma A of section 1, chapter I, and the fact that $\tfrac{\partial}{\partial x_0} A = (\tfrac{\partial}{\partial x_0} C^\tau) C + C^\tau \tfrac{\partial}{\partial x_0} C$ we see that (1.6), (1.7), and (**) of Theorem 1.10 hold for A. Also,

$$B = \lambda (\nabla \tfrac{\partial}{\partial x_0} P_{\gamma x_0} \psi)^\tau + \lambda \tfrac{\partial}{\partial t} P_{\gamma x_0} \psi \, e_0$$

so again from Lemma A we find that (1.5) and (**) are valid for B. To verify (1.9) and (***) observe that $\tfrac{\partial}{\partial x_0} A_{ij} = 0$ for $1 \le i, j \le n-1$ and

$$\tfrac{\partial}{\partial x_0} A_{ij} = \sum_{l=0}^{n-1} \langle e_l^{ij}, \tfrac{\partial}{\partial x_l} f_l^{ij} \rangle$$

where $f_l^{ij} = \tfrac{\partial}{\partial x_0} P_{\gamma x_0} \psi$ when either $i = 0$ or $j = 0$ and

$$e_0^{00} = -2\lambda^3 [\, 1 + \sum_{k=1}^{n-1} (\tfrac{\partial}{\partial x_k} P_{\gamma x_0} \psi)^2 \,]$$

(2.31)
$$e_l^{00} = 2\lambda^2 \tfrac{\partial}{\partial x_l} P_{\gamma x_0} \psi, \ 1 \le l \le n-1,$$

$$e_0^{i0} = \lambda^2 \tfrac{\partial}{\partial x_i} P_{\gamma x_0} \psi, \ 1 \le i \le n-1,$$

$$e_l^{i0} = -\lambda \delta_{il} \text{ for } 1 \le i, l \le n-1.$$

In the above display, $\delta_{il} = 1$ when $i = l$ and $\delta_{il} = 0$ otherwise, is the Kronecker δ. Using symmetry of A, Lemma A, (2.30), and (2.31) it is easily checked that (1.9) and (***) holds. Thus A, B as above satisfy the hypotheses of Theorem 1.10 so we can apply this theorem to conclude that

$$\| \tfrac{d\hat{\omega}}{dyds}(d, x, t + 2d^2, \cdot) \|_{\alpha_p(Q_d(x,t))} < c^+ < \infty .$$

Since ω is known to be doubling (see[FGS], [H]) it follows that α_p can be replaced by β_p in the above inequality. The proof of Corollary 1.12 is now complete. $\square$

3. LEMMAS ON PARABOLIC MEASURE

In this section we prove several lemmas on parabolic measure which are well known for parabolic measures satisfying the conclusion of Lemmas 3.14 and 3.38 of chapter I . In fact most of our effort will be devoted to overcoming our inability to prove that certain parabolic measures in chapter III are doubling. We first examine the implications of the conclusion of Lemma 2.1. That is, suppose the continuous Dirichlet problem corresponding to (1.1) and some A_1, B_1 satisfying (1.2)-(1.4) always has a unique solution. Let ω_1 be parabolic measure corresponding to A_1, B_1, and assume there exist $\eta_0, \eta_1 > 0$ such that whenever $Q_r(y, s) \subset Q_d(x, t), E \subset Q_r(y, s)$, and E is Borel, we have

(3.1) $|E|/|Q_r(y, s)| \geq 1 - \eta_0 \Rightarrow \omega_1(r, y, s + 2r^2, E) \geq \eta_1.$

In the proof of Theorem 1.10 we shall need the following analogue of Lemma 3.9 in chapter I.

Lemma 3.2. *Let A_1, B_1 satisfy (1.2)-(1.4) and assume that the continuous Dirichlet problem corresponding to (1.1), A_1, B_1 always has a unique solution. Let ω_1 be parabolic measure for (1.1), A_1, B_1, and suppose that (3.1) holds for ω_1 whenever $E = Q_r(y, s)$. Let $Q_{2r}(y, s) \subset Q_d(x, t)$ and let u_1 be a weak solution to (1.1) in $(0, 2r) \times Q_{2r}(y, s)$ corresponding to A_1, B_1. If u_1 vanishes continuously on $Q_{2r}(y, s)$, then there exists $c = c(\gamma_1, M, \eta_0, \eta_1, n)$ and $\alpha = \alpha(\gamma_1, M, \eta_0, \eta_1, n), 0 < \alpha < 1 \leq c < \infty$, such that*

$$u_1(Z, \tau) \leq c \, (z_0/r)^\alpha \max_{(0, r) \times Q_r(y, s)} u_1$$

whenever $(Z, \tau) \in (0, r/2) \times Q_{r/2}(y, s)$. If $u_1 \geq 0$ in $(0, 2r) \times Q_{2r}(y, s)$, then there exists $\tilde{c} = \tilde{c}(\gamma_1, M, \eta_0, \eta_1 n)$ such that for (Z, τ) as above,

$$u_1(Z, \tau) \leq \tilde{c} \, (z_0/r)^\alpha \, u_1(r, y, s + 2r^2).$$

Proof: To prove Lemma 3.2 let $0 < \rho < r/4$ and $(z, \tau) \in Q_{r/2}(y, s)$. Set $G_j = \bar{Q}_{2^{-j}\rho}(z, \tau) \setminus Q_{2^{-(j+1)}\rho}(z, \tau)$, for $j = 0, 1, \ldots,$. In (3.1) we take $E = Q_r(y, s)$ and use this implication repeatedly with $Q_r(y, s)$ as certain subrectangles of G_j, as well as Harnack's inequality, to concude the existence of $c^* \geq 1$ such that

(3.3) $c^*\omega(\cdot, G_j) \geq 1$

on

$$[0, 2^{-j}\rho] \times \partial Q_{2^{-j-1/2}\rho}(z, \tau) \ \cup \ \{2^{-j}\rho\} \times Q_{2^{-j-1/2}\rho}(z, \tau) \text{ for } j = 0, 1, \ldots, .$$

Here the boundary of the above set is taken with respect to R^n. Using the maximum principle for solutions to (1.1) (see the remark after Lemma 3.38 of chapter I) and the definition of ω_1 it follows that

(3.4) $$\omega_1(\cdot, \bigcup_{i=1}^{j-1} G_i) \leq c^* \omega_1(\cdot, G_j)$$

in

$$(0, 2^{-j}\rho) \times Q_{2^{-j-1/2}\rho}(z, \tau) \text{ for } j = 1, 2, \ldots,.$$

Clearly this inequality implies

$$\omega_1\Big(\cdot, \bigcup_{i=1}^{j-1} G_i\Big) \leq \frac{c^*}{c^* + 1} \omega_1\Big(\cdot, \bigcup_{i=1}^{j} G_i\Big) \text{ in } (0, 2^{-j}\rho) \times Q_{2^{-j-1/2}\rho}(z, \tau), \, j = 1, 2, \ldots,.$$

Iterating the above inequality starting with $j = 1$ we see that if $\beta = c^*/(c^*+1) < 1$, then

$$(3.5) \qquad \omega_1(\cdot, G_0) \leq \beta^j \text{ in } (0, 2^{-j}\rho) \times Q_{2^{-j-1/2}\rho}(z, \tau), \, j = 1, \ldots,.$$

To conclude the proof of Lemma 3.2 we observe first from the maximum principle for (1.1), the definition of ω_1, and (3.3) that

$$u_1 \leq c^* \max_{(0,r) \times Q_r(y,s)} u_1 \; \omega_1(\cdot, G_0) \text{ in } (0, \rho/2) \times Q_{\rho/2}(z, \tau),$$

and second from (3.5) with $\rho = r/4$ that the first part of Lemma 3.2 is valid. The second part of Lemma 3.2 for $u_1 \geq 0$ follows from the first part of this lemma and Harnack's inequality by a standard argument mentioned in the proof of Lemma 3.14 of chapter I. □

Next we prove

Lemma 3.6. *Let A_1, B_1 satisfy (1.2)-(1.4) and assume that the continuous Dirichlet problem corresponding to A_1, B_1 always has a unique solution. Let ω_1 be parabolic measure for (1.1), A_1, B_1, and suppose that (3.1) holds for ω_1. Then for some $p, 1 < p < \infty$, we have $\frac{d\omega_1}{dyds}(d, x, t + 2d^2, \cdot) \in \alpha_p(Q_d(x, t))$ with*

$$\| \tfrac{d\omega_1}{dyds}(d, x, t + 2d^2, \cdot) \|_{\alpha_p(Q_d(x,t))} \leq c(\gamma_1, M, \eta_0, \eta_1, n) < \infty.$$

Proof: We remark that Lemma 3.6 would be an immediate consequence of the results in [CF] if we knew that ω_1 was a doubling measure. To begin the proof, we claim for given $\epsilon > 0$ that there exists $\hat{c} = \hat{c}(\epsilon, \gamma_1, M, \eta_0, \eta_1, n), c_- = c_-(n) \geq 1$ such that the following statement is true whenever $Q_\rho(z, \tau) \subset Q_d(x, t)$. If $E \subset Q_\rho(z, \tau)$ is a Borel set and $|E|/|Q_\rho(z, \tau)| \geq 1 - \eta_0/c_-$, then

$$(3.7)$$
$$\omega_1(\cdot, Q_{\rho/2}(z, \tau)) \leq \epsilon \omega_1(\cdot, Q_\rho(z, \tau)) + \hat{c}\omega_1(\cdot, E) \quad \text{in } U \setminus [[0, \rho] \times \bar{Q}_\rho(z, \tau)].$$

To prove this claim we first show that if $G_0(\eta) = Q_{(1+\eta)\rho/2}(z, \tau) \setminus Q_{(1-\eta)\rho/2}(z, \tau)$, then there exists c_+, θ depending on $\eta_0, \eta_1, \gamma, M, n$ with

$$(3.8)$$
$$\omega_1(\cdot, G_0(\eta)) \leq c_+ \eta^\theta \; \omega_1(\cdot, Q_\rho(z, \tau)) \text{ in } U \setminus [[0, \rho] \times \bar{Q}_\rho(z, \tau)] \text{ for } 0 < \eta < 1/10^{10}.$$

The proof of (3.8) is similar to the proof of (3.5). Let

$$\hat{G}_j(\eta) = Q_{(1+2^{j+1}\eta)\rho/2}(z, \tau) \setminus \bar{Q}_{(1+2^j\eta)\rho/2}(z, \tau),$$

$$\tilde{G}_j(\eta) = Q_{(1-2^j\eta)\rho/2}(z, \tau) \setminus \bar{Q}_{(1-2^{j+1}\eta)\rho/2}(z, \tau),$$

$$G_j(\eta) = \hat{G}_j(\eta) \cup \tilde{G}_j(\eta),$$

for $j = 0, 1, \ldots,$ and $2^{j+1}\eta \leq 1$. We apply (3.1) with $Q_r(y, s) =$ certain subrectanges $\subset G_j(\eta)$ and with $E = Q_r(y, s)$. We get that (3.3) holds with G_j replaced by $G_j(\eta)$ on the union of $[0, 2^j\eta\rho] \times \partial Q_{(1\pm 2^{j+1/2}\eta)\rho/2}(z, \tau)$ and

$$\{2^j\eta\rho\} \times [Q_{(1+2^{j+1/2}\eta)\rho/2}(z, \tau) \setminus Q_{(1-2^{j+1/2}\eta)\rho/2}(z, \tau)]$$

for $j = 1, \ldots,$. Using (3.3) and the maximum principle for solutions to (1.1) (see the remark after Lemma 3.33 of chapter I), we find that (3.4) holds in

$$U \setminus \{[0, 2^j\eta\rho] \times (\bar{Q}_{(1+2^{j+1/2}\eta)\rho/2}(z, \tau) \setminus Q_{(1-2^{j+1/2}\eta)\rho/2}(z, \tau))\}$$

provided $2^j\eta \leq 1/3$. Iterating (3.4) we obtain (3.8).

Next we observe from (3.1) that for some $c_0 \geq 1$ we have

$$c_0\,\omega_1(\cdot, G_0(\eta)) \geq 1 \text{ on } (0, \eta\rho) \times \partial Q_{\rho/2}(z, \tau).$$

To prove claim (3.7) we choose η, $0 < \eta \leq 2^{-3}$, to be the largest number such that

(3.9) $c_0\,\omega_1(\cdot, G_0(\eta)) \leq \epsilon\,\omega_1(\cdot, Q_\rho(z, \tau)) \text{ in } U \setminus [[0, \rho] \times \bar{Q}_\rho(z, \tau)].$

With η now fixed we use the bisection method, a weak type argument, and (3.1) to find that if $c_-(n)$ is large enough and $r = 2^{-j}\rho$, $j = 5, 6, \ldots$, then there exists $Q_r(y, s) \subset Q_{\rho/4}(z, \tau - \rho^2/2)$ with $|E \cap Q_r(y, s)| \geq (1 - \eta_0)|Q_r(y, s)|$. Choosing $r = 2^{-10}\rho$ and using (3.1) with E replaced by $E \cap Q_r(y, s)$, as well as Harnack's inequality, we find the existence of $\hat{c}$ as in (3.7) with

$$\hat{c}\,\omega_1(\cdot, E) \geq 1$$

on

$$[\eta\rho, \rho] \times \partial Q_{\rho/2}(z, \tau) \cup \{\rho\} \times Q_{\rho/2}(z, \tau).$$

From the maximum principle noted above we conclude first that

$$\omega_1(\cdot, Q_{\rho/2}(z, \tau)) \leq c_0\,\omega_1(\cdot, G_0(\eta)) + \hat{c}\,\omega_1(\cdot, E) \text{ in } U \setminus [[0, \rho] \times \bar{Q}_{\rho/2}(z, \tau)]$$

and second from this inequality, as well as (3.9), that claim (3.7) is valid. We also claim there exists $\beta > 0, 0 < \beta < 1/2$, and $\tilde{c} \geq 2$, depending on $\gamma_1, M, \eta_0, \eta_1, n$ such that if $Q_r(y, s) \subset Q_{3\rho/2}(z, \tau) \subset Q_{2\rho}(z, \tau) \subset Q_d(x, t)$, then

(3.10)
$$\tilde{c}^{-1}\,(r/d)^{1/\beta} \leq \omega_1(d, x, t + 2d^2, Q_r(y, s)) \leq \tilde{c}\,(r/\rho)^\beta\,\omega_1(d, x, t + 2d^2, Q_{2\rho}(z, \tau)).$$

The lefthand side of (3.10) follows from (3.1) and Harnack's inequality. The righthand side of (3.10) is proved by an argument similar to the one used in proving (3.8).

Armed with the above claims we are ready to show that ω_1 restricted to $Q_d(x, t)$ is absolutely continuous with respect to Lebesgue measure. To do this first observe for some $K \geq 2, 0 < K < \infty$, that

(3.11) $$\liminf_{r \to 0} \frac{\omega_1(d, x, t + 2d^2, Q_r(y, s))}{\omega_1(d, x, t + 2d^2, Q_{r/2}(y, s))} \leq K$$

whenever $(y, s) \in Q_d(x, t)$. Indeed if the above inequality were false for large K, we could use the lefthand inequality in (3.10) and iteration to get a contradiction. Second we observe from a standard argument (using the Besicovitch covering lemma) that for $\omega_1(d, x, t + 2d^2, \cdot)$ almost every $(y, s) \in F$ Borel $\subset Q_d(x, t)$ we have

(3.12) $$\lim_{r \to 0} \frac{\omega_1(d, x, t + 2d^2, Q_r(y, s) \setminus F)}{\omega_1(d, x, t + 2d^2, Q_r(y, s))} \equiv 0.$$

Now if $\omega_1(d, x, t + 2d^2, \cdot)$ were not absolutely continuous with respect to Lebesgue measure on $Q_d(x, t)$, then for some F Borel $\subset Q_d(x, t)$ we would have $|F| = 0$ and $\omega_1(d, x, t + 2d^2, F) > 0$. Choose $(y, s) \in F$ so that both (3.11), (3.12) hold. To get a contradiction we note from (3.11), (3.7), with $Q_\rho(z, \tau)$ replaced by $Q_r(y, s)$ and E by $Q_r(y, s) \setminus F$ that if ϵ is small enough (depending on K in (3.11)) we have

$$\omega_1(d, x, t + 2d^2, Q_r(y, s)) \leq \xi \,\omega_1(d, x, t + 2d^2, Q_r(y, s) \setminus F)$$

for some arbitrary small $r > 0$ and some $\xi > 0$ independent of r. Clearly this inequality contradicts (3.12). Thus by the Radon-Nikodym theorem, whenever $G \subset Q_d(x, t)$ is a Borel set, we have

$$\omega_1(d, x, t + 2d^2, G) = \int_G f \, dz d\tau$$

for some Borel measurable $f \geq 0$ with $\|f\|_{L^1(Q_d(x,t))} \leq 1$.

To continue we use an argument essentially due to Gehring (see [G], [Gi], and [CF]). Fix $Q_{2r'}(y', s') \subset Q_d(x, t)$. Given

$$(3.13) \qquad \lambda > (1000)^{100n} \, |Q_{2r'}(y', s')|^{-1} \int_{Q_{2r'}(y', s')} f \, dy ds \; = \; \lambda_0$$

suppose $(y, s) \in Q_{r'}(y', s')$ is a point of Lebesgue density 1 of

$$F(\lambda) = \{(z, \tau) \in Q_d(x, t) : f(z, \tau) > \lambda\}.$$

Using (3.13) and continuity of the integral we see there exists r, $0 < r < r'/1000$, such that

$$(3.14) \quad
\begin{aligned}
&(a) \quad \lambda = |Q_{10r}(y, s)|^{-1} \int_{Q_{10r}(y, s)} f \, dz d\tau, \\[2mm]
&(b) \quad |Q_\rho(y, s)|^{-1} \int_{Q_\rho(y, s)} f \, dz d\tau > \lambda \text{ for } 0 < \rho < 10r.
\end{aligned}$$

We note from (3.14)(b) that

$$(3.15) \qquad (20)^{n+1} \, \omega_1(d, x, t + 2d^2, Q_{r/2}(y, s)) \; \geq \; \omega_1(d, x, t + 2d^2, Q_{10r}(y, s)).$$

Set

$$E(\delta\lambda) = \{(z, \tau) \in Q_d(x, t) : f(z, \tau) \leq \delta\lambda\}$$

and suppose that $|E(\delta\lambda) \cap Q_r(y, s)| \geq (1 - \eta_0/c_-(n)) \, |Q_r(y, s)|$. Then from (3.15), (3.7) with ρ, z, τ replaced by r, y, s, we see for ϵ sufficiently small that

$$\omega_1(d, x, t + 2d^2, Q_r(y, s)) \leq c \,\omega_1(d, x, t + 2d^2, E(\delta\lambda) \cap Q_r(y, s))$$

where c depends only on $\gamma_1, M, \eta_0, \eta_1, n$. Dividing this inequality by $|Q_r(y, s)|$ we deduce from simple estimates using (3.14)(a), (3.15), that for some $c' \geq 2$,

$$1 \leq c' \delta.$$

If $\delta_0 = \frac{1}{2c'}$, where c' is the above constant, then from the above inequality and (3.14)(a) we see that

$$(3.16) \qquad |Q_r(y, s) \cap F(\delta_0\lambda)| \geq (\eta_0/c_-) \, |Q_r(y, s)|.$$

Using a well known covering argument we get a sequence $\{Q_{10r_i}(y_i, s_i)\}$ of rectangles for which (3.14)-(3.16) holds and also

$$(+) \qquad \{(y_i, s_i)\} \subset F(\lambda) \cap Q_{r'}(y', s')$$

$$(3.17) \qquad (++) \qquad \left| (F(\lambda) \cap Q_{r'}(y', s')) \setminus \bigcup_i Q_{10r_i}(y_i, s_i) \right| = 0,$$

$$(+++) \quad Q_{r_i}(y_i, s_i) \cap Q_{r_j}(y_j, s_j) = \emptyset \text{ when } i \neq j.$$

Let H_1 be the subfamily of rectangles $Q \in \{Q_{r_j}(y_j, s_j)\}$ with $Q \setminus Q_{r'}(y', s') \neq \emptyset$ and put $H_2 = \{Q_{r_j}(y_j, s_j)\} \setminus H_1$. If $Q_r(y, s) \in H_1$ we note from (3.10) and (3.14)(b) that

$$\lambda |Q_r(y, s)| \leq \int_{Q_r(y, s)} f \, dz d\tau \leq c(r/r')^\beta \omega_1(d, x, t + 2d^2, Q_{2r'}(y', s')).$$

Solving this inequality for r and using the definition of λ_0 in (3.13) we get

$$(3.18) \qquad r \leq c(n) \, r' \, (\lambda_0/\lambda)^{1/(n+1-\beta)} = \eta r'/(100n).$$

Hence $\bigcup_{Q \in H_1} Q \subset Q_{r'(1+\eta)}(y', s') \setminus Q_{r'(1-\eta)}(y', s')$ and we can argue as in the proof of (3.7) to conclude as in (3.8) that

$$(3.19) \qquad \bigcup_{Q \in H_1} \omega_1(d, x, t + 2d^2, \bar{Q}) \leq c \, (\lambda_0/\lambda)^{\theta_1} \, \omega_1(d, x, t + 2d^2, Q_{2r'}(y', s'))$$

where $c = c(\gamma_1, M, \eta_0, \eta_1, n)$ and $\theta_1 = \theta/(n+1-\beta)$. Using (3.14)-(3.19) we obtain

$$(3.20)$$
$$\int_{F(\lambda) \cap Q_{r'}(y', s')} f \, dz d\tau \leq \sum_i \int_{Q_{10r_i}(y_i, s_i)} f \, dz \, d\tau$$

$$\leq (20)^{n+1} \sum_i \int_{Q_{r_i}(y_i, s_i)} f \, dz d\tau = (20)^{n+1} \Big[\sum_{Q \in H_1} \int_Q f \, dz d\tau + \sum_{Q \in H_2} \int_Q f \, dz d\tau \Big]$$

$$\leq c \, (\lambda_0/\lambda)^{\theta_1} \, \omega_1(d, x, t + 2d^2, Q_{2r'}(y', s')) + c \lambda \sum_{Q \in H_2} |Q|$$

$$\leq c(\lambda_0/\lambda)^{\theta_1} \, \omega_1(d, x, t + 2d^2, Q_{2r'}(y', s')) + c \lambda \sum_{Q \in H_2} |Q \cap F(\delta_0 \lambda)|$$

$$\leq c(\lambda_0/\lambda)^{\theta_1} \, \omega_1(d, x, t + 2d^2, Q_{2r'}(y', s')) + c \lambda \, |F(\delta_0 \lambda) \cap Q_{r'}(y', s')|.$$

This inequality implies (see [Gi, ch 5]) the existence of $\theta_2, 0 < \theta_2 < \theta_1/2$, such that
$$(3.21)$$

$$|Q_{r'}(y', s')|^{-1} \int_{Q_{r'}(y' s')} f^{1+\theta_2} \, dz d\tau \leq c \left(|Q_{2r'}(y', s')|^{-1} \int_{Q_{2r'}(y', s')} f \, dz d\tau \right)^{1+\theta_2}.$$

To see that (3.20) implies (3.21) we multiply (3.20) by $\lambda^{-1+\theta_2}$ and integrate from λ_0 to ∞. After a careful limiting argument we deduce for sufficiently small θ_2 that the integral corresponding to the last term in (3.20) can be absorbed into the integral corresponding to the lefthandside of this equation. The resulting integrals are equivalent to (3.21). Since $Q_{2r'}(y', s') \subset Q_d(x, t)$ is arbitrary we conclude that Lemma 3.6 is true. $\square$

To set the stage for our next lemma suppose $K \subset Q_d(x, t)$ is a nonempty closed set and $0 \leq \psi \leq 10^{-2n} \sigma(\cdot, K)$, where $\sigma, \psi : R^n \to R$ are as in (2.19) and (2.25), respectively. Let $\rho : U \to U$ be as in the display following (2.25) and $\gamma > 0$ as in (2.26). Let u_1 be a solution to (1.1) in U corresponding to A_1, B_1 satisfying (1.2)-(1.4) with constants γ_1, M. Then from (2.25)-(2.28) and Lemma A in section 1 we deduce that $u_2 = u_1 \circ \rho$ is a weak solution to (1.1) corresponding to some A_2, B_2. Also A_2, B_2 satisfy (1.2)-(1.4) with constants depending only on n, the constants for A_1, B_1, and a_1 in (2.25). With this notation we prove

Lemma 3.22. *Let* $K, Q_d(x, t), A_1, B_1, A_2, B_2$ *be as above. Suppose also that the Dirichlet problem for* A_1, B_1 *and* A_2, B_2 *always has a unique solution. Let* ω_1, ω_2 *be the corresponding parabolic measures and assume that (3.1) is valid for* ω_1 *while the conclusion of Lemma 3.37 holds with* $\omega = \omega_2$. *Then there exists* $\alpha > 0, \tilde{c} \geq 1$, *depending only on* $n, \gamma_1, M, a_1, \gamma, \eta_0, \eta_1$, *and the constant in Lemma 3.37 of chapter I, such that*

$$\tilde{c}\,\omega_2(d, x, t + 2d^2, K) \geq \left(\frac{|K|}{|Q_d(x, t)|} \right)^\alpha .$$

Proof: We remark that if we knew ω_1 were a doubling measure, then we could use the same argument as in [DJK] to get Lemma 3.22. Instead we use an argument based on the Calderón - Zygmund decomposition and an elaboration of the argument following (2.23). To begin the proof we first note from Lemma 3.6 that if $Q_{2r}(y, s) \subset Q_d(x, t)$ and $f = \frac{d\omega_1}{dy\,ds}(2r, y, s + 4r^2, \cdot)$, then there exists $p > 1$ such that $\|f\|_{\alpha_p(Q_{2r}(y,s))} < \infty$. Hence if G is a Borel subset of $Q_r(y, s)$, then

(3.23)

$$\omega_1(2r, y, s + 4r^2, Q_r(y, s) \setminus G) = \int_{Q_r(y,s)\setminus G} f \, dz d\tau$$

$$\leq |Q_r(y, s) \setminus G|^{(p-1)/p} \left(\int_{Q_r(y,s)} f^p dz d\tau \right)^{1/p} \leq c \left(\frac{|Q_r(y, s) \setminus G|}{|Q_r(y, s)|} \right)^{(p-1)/p},$$

where c depends on the constant in Lemma 3.6. Next we use a construction of Whitney to write $Q_d(x, t) \setminus K = \bigcup \bar{Q}_i$, where $\{Q_i\}$ are parabolic rectangles with disjoint interiors and side length in the space direction proportional to their parabolic distance from K. Also we choose these rectangles from the family of all rectangles obtained by bisecting the sides of $Q_d(x, t)$ into rectangles of side length $2^{1-m}d$ in the space direction and side length $2 \cdot 4^{-m}d^2$ in the time direction for $m = 1, 2, \ldots,$. Let $\hat{K} \subset Q_d(x, t)$ be the union of K and certain of the above closed Whitney rectangles. We use (3.23) to show the existence of $\theta_1, \theta_2, 0 < \theta_1, \theta_2 < 1/2$, having the same dependence as $\tilde{c}$ in Lemma 3.22, such that if $Q_r(y, s) \subset Q_d(x, t)$, then

(3.24) $$\frac{|\hat{K} \cap Q_r(y, s)|}{|Q_r(y, s)|} \geq 1 - \theta_1 \Rightarrow \omega_2(r, y, s + 2r^2, \hat{K} \cap Q_r(y, s)) \geq \theta_2.$$

Thus (3.24) implies Lemma 3.22 when $K = \hat{K}$ is most of $Q_d(x, t)$.

Let $\mathcal{F}$ be the family of all open Whitney rectangles, Q_i, such that $\bar{Q}_i \cap \bar{Q}_{r/3}(y, s)$ $\neq \emptyset$. We note that if $Q_i \in \mathcal{F}$, then from the geometry of Whitney rectangles, there exists $Q_{i'} \in \mathcal{F}$ with $\bar{Q}_i \cap \bar{Q}_{i'} \neq \emptyset$ and

$$c \, |Q_{i'} \cap Q_{r/3}(y, s)| \geq \min\{|Q_i|, |Q_r(y, s)|\},$$

where $c = c(n)$ depends on the ratio of the side lengths of neighboring Whitney rectangles. Let $\mathcal{F}_1$ be the subfamily of $\mathcal{F}$ consisting of rectangles Q_i with $\bar{Q}_{i'} \subset \hat{K}$ and let $\mathcal{F}_2$ be the set of all rectangles in $\mathcal{F}$ which are not in $\mathcal{F}_1$. From the doubling property of ω_2 (Lemma 3.37 (β) of chapter I) and the above note we find that

$$(3.25)$$
$$\sum_{Q_i \in \mathcal{F}_1} \omega_2(r/2, y, s + r^2/2, \bar{Q}_i \cap Q_{r/3}(y, s)) \leq c \, \omega_2(r/2, s + r^2/2, \hat{K} \cap Q_{r/3}(y, s))$$

where c has the same dependence as the constant in Lemma 3.37. If $Q_i \in \mathcal{F}_2$ and $r_1 = (r/3)(1 + c \theta_1^{1/(n+1)})$, then for c large enough (depending only on n and the ratio of the side length of Q_i to its distance from K), we have

$$(3.26) \qquad \bar{Q}_{i'} \subset Q_{r_1}(y, s) \setminus \hat{K}.$$

Let ρ be defined as above Lemma 3.22 and set $\omega^*(z, \tau, E) = \omega_1(\rho(z, \tau), E)$, $(z, \tau) \in \bar{U}$, whenever $E \subset R^n$ is a Borel set. Note that $\omega^*(\cdot, E)$ satisfies (1.1) relative to A_2, B_2. If $Q_i = Q_{\hat{r}}(\hat{y}, \hat{s}) \in \mathcal{F}_2$, then from (3.1), Harnack's inequality, and the fact that $0 \leq \psi \leq 10^{-2n}\sigma$, we deduce first that $c \omega^*(\cdot, Q_{\hat{r}}(\hat{y}, \hat{s})) \geq 1$ on $Q_{\hat{r}/4}(\hat{y}, \hat{s})$ and second from Harnack's inequality, Lemma 3.37 (β) of chapter I, the definition of ω_2, and (3.26) that

$$(3.27) \qquad \omega_2(r, y, s + 2r^2, Q_{i'}) \leq c^* \omega^*(r, y, s + 2r^2, Q_{i'}).$$

Let $G = \bigcup_{Q_i \in \mathcal{F}_2} \bar{Q}_i$, $G' = \bigcup_{Q_i \in \mathcal{F}_2} \bar{Q}_{i'}$. Then from (3.23), (3.26), (3.27) and Lemma 3.37 $(\alpha), (\beta)$, we find for $\theta_1 > 0$ small enough that

$$\omega_2(r/2, y, s + r^2/2, G) \leq c \, \omega_2(r/2, y, s + r^2/2, G')$$

$$\leq c c^* \omega^*(r/2, y, s + r^2/2, G') \leq \tfrac{1}{2} \omega_2(r/2, y, s + r^2/2, \bar{Q}_{r/3}(y, s))$$

This inequality and (3.25) imply that

$$(3.28) \quad \omega_2(r/2, y, s + r^2/2, \bar{Q}_{r/3}(y, s)) \leq c \, \omega_2(r/2, y, s + r^2/2, \hat{K} \cap Q_{r/3}(y, s)).$$

Using, (3.28), Lemma 3.37 (α), and Harnack's inequality we see that (3.24) is true.

Next let $\phi = 10^{-(n+1)} \theta_1$, $\theta = \theta_2/c_1$, where c_1 is a large positive constant to be specified later. We shall show there exists a positive integer m and a sequence of

Borel sets $\{K_j\}_0^m$ such that

(a) \qquad $K_0 = K, K_m = \bar{Q}_d(x,t)$, and $K_i \subset K_j$ for $i < j$,

(b) \qquad If $m > 1$, then $(1 - \phi)|K_i| \geq |K_{i-1}|$, for $1 \leq i \leq m - 1$,

(3.29) \quad (c) \qquad $|K_{m-1}|/|Q_d(x,t)| \leq 1 - \theta_1$,

(d) \qquad $\omega_2(d, x, t + 2d^2, K_{i-1}) \geq \theta\, \omega_2(d, x, t + 2d^2, K_i), 1 \leq i \leq m,$

(e) \qquad $K_i, 1 \leq i \leq m$, is the union of K and certain Q_j.

We note that (3.29) implies

$$|K|/|Q_d(x,t)| \leq (1 - \phi)^{m-1} \text{ and}$$

$$\omega_2(d, x, t + 2d^2, K) \geq \theta^m\, \omega_2(d, x, t + 2d^2, Q_d(x,t)).$$

Clearly these inequalities and Lemma 3.37 (α) imply Lemma 3.22 with $\alpha = \frac{\ln \theta}{\ln(1-\phi)}$. Thus to complete the proof of Lemma 3.22 we need only prove (3.29). To do this we proceed by induction. Suppose, for some nonnegative integer l that $K_0, \ldots, K_l$ have been constructed satsifying $(a) - (e)$ in (3.29). If $|K_l|/|Q_d(x,t)| \geq 1 - \theta_1$ we put $m = l + 1$. Using (3.24) with $\hat{K}$ replaced by K_l, we see that $\omega_2(d, x, t + 2d^2, K_l) \geq \theta\, \omega_2(d, x, t + 2d^2, Q_d(x,t))$. Thus (3.29) is valid in this case. If $|K_l|/|Q_d(x,t)| < 1 - \theta_1$, we use the method of Calderón-Zygmund to get sequences $\{L_j\}, \{L'_j\}$ of open parabolic rectangles satisfying

(i) \qquad $(1 - \theta_1)\,|L'_j| \leq |K_l \cap L'_j|,$

(ii) \qquad $|K_l \cap L_j| < (1 - \theta_1)|L_j|,$

(3.30) \quad (iii) \qquad $|K_l \setminus (\bigcup L'_j)| = 0,$

(iv) \qquad $L'_j \subset L_j$ and $|L_j| = 2^{n+1}\,|L'_j|,$

(v) \qquad $L'_i \cap L'_j = \emptyset$ for $i \neq j$.

Let

$$K_{l+1} = K_l \bigcup \{\bar{Q}_i : \bar{Q}_i \cap L_j \neq \emptyset \text{ for some } j\}.$$

We first show that (3.29)(b) holds with $i = l + 1$. To do this we use a wellknown covering argument to get a subsequence $\{L^*_j\}$ of $\{L_j\}$ consisting of disjoint parabolic rectangles with

$$10^{n+1} \sum |L^*_j| \geq |\bigcup L_j|.$$

Let $L^* = \bigcup L_j^*$, $L = \bigcup L_j$. Then from (3.30) (ii) we see that $|K_l \cap L^*| \le (1 - \theta_1)|L^*|$. From this inequality, the above inequality, and (3.30) (iii) we find

$$|K_l| = |K_l \cap L^*| + |K_l \setminus L^*|$$

$$\le (1 - \theta_1)|L^*| + |L \setminus L^*|$$

$$\le |L| - \theta_1|L^*|$$

$$\le (1 - \phi)|L|$$

$$\le (1 - \phi)|K_{l+1}|.$$

Thus (3.29) (b) is valid with $i = l + 1$.

Next we show that (3.29) (d) is true with $i = l + 1$. Let $\mathcal{H}_1$ be the family of all $Q_i \subset K_{l+1} \setminus K_l$ for which there exists $Q_{i'} \subset K_l$ with $\bar{Q}_i \cap \bar{Q}_{i'} \ne \emptyset$. Using Lemma 3.37 (β) of chapter I, once again, we see that

$$(3.31) \qquad \sum_{Q_i \in \mathcal{H}_1} \omega_2(d, x, t + 2d^2, \bar{Q}_i) \le c\,\omega_2(d, x, t + 2d^2, K_l).$$

Put $\mathcal{H}_2 = \{Q_i \subset K_{l+1} \setminus K_l : Q_i \notin \mathcal{H}_1\}$. If $Q_i \in \mathcal{H}_2$, then from (3.30) $(i), (ii), (iv)$, and the definition of K_{l+1} we deduce the existence of $j = j(i)$ and $c_+(n) \ge 1$ such that if $L_j = Q_{\hat{r}}(\hat{y}, \hat{s})$, then $Q_i \subset Q_{c_+ \hat{r}}(\hat{y}, \hat{s})$. From this deduction, Lemma 3.37 (β), and (3.30) (iv) we conclude for a given L_j that if L_j'' is the union of all $\bar{Q}_i \in \mathcal{H}_2$ such that $j = j(i)$, then -

$$(3.32) \quad \omega_2(d, x, t + 2d^2, L_j'') \le c\omega_2(d, x, t + 2d^2, L_j) \le c^2\,\omega_2(d, x, t + 2d^2, L_j').$$

Now from (3.24) and (3.30) (i) we find also for $L_j' = Q_{r'}(y', s')$ that

$$c'\omega_2(r', y', s' + 2(r')^2, K_l \cap L_j') \ge 1.$$

From Lemma 3.37 (γ) and the above inequality it follows that

$$c''\omega_2(d, x, t + 2d^2, K_l \cap L_j') \ge \omega_2(d, x, t + 2d^2, L_j').$$

Using this inequality in (3.32) and (3.30)(v) we get

$$\omega_2(d, x, t + 2d^2, \bigcup_{Q_i \in \mathcal{H}_2} \bar{Q}_i) \le c\,\omega_2(d, x, t + 2d^2, K_l).$$

Clearly this inequality and (3.31) yield (3.29)(d). (3.29)(e) is included in the definition of K_{l+1}. By induction we obtain (3.29). From the remark after (3.29) we see that Lemma 3.22 is true. $\square$

Finally in this section suppose that $K = \bigcup_{i=1}^N \bar{Q}_{r_i}(y_i, s_i)$, $K' = \bigcup_{i=1}^N \bar{Q}_{2r_i}(y_i, s_i)$, and $K \subset K' \subset Q_d(x, t)$. Also assume that $\bar{Q}_{2r_i}(y_i, s_i) \cap \bar{Q}_{2r_j}(y_j, s_j) = \emptyset$ when $i \ne j$. Let ψ, σ be as in (2.19), (2.25), respectively and suppose that $0 \le \psi \le 10^{-2n}\sigma(\cdot, K)$ on $Q_d(x, t) \setminus K'$ while $0 \le \psi \le 10^{-2n}r_i$ on $\bar{Q}_{2r_i}(y_i, s_i)$ for $1 \le i \le N < \infty$. Define ρ relative to ψ as in the display after (2.25). We close this section with

Lemma 3.33. *Let ω_1 be as in Lemma 3.22 and define ω_2 relative to ω_1, ρ as in this lemma. Suppose that (3.1) is valid for ω_1 while the conclusion of Lemma 3.37*

of chapter I holds with $\omega = \omega_2$. Then there exists $\alpha > 0, \tilde{c} \geq 1$, depending only on $n, \gamma_1, M, a_1, a_2, \gamma, \eta_0, \eta_1$, and the constant in Lemma 3.37, such that

$$\tilde{c}\,\omega_2(d, x, t + 2d^2, K) \geq \left(\frac{|K \cap Q_d(x,t)|}{|Q_d(x,t)|} \right)^\alpha.$$

Proof: Note that Lemma 3.33 does not follow directly from Lemma 3.22 because of our relaxed assumptions on ψ. However if $\{\bar{Q}_i\}$ is a Whitney decomposition of $Q_d(x,t) \setminus K$, as previously, then we can repeat verbatim the argument in this lemma to get first (3.24) with K replaced by

$$K'' = K' \cup \{(z,\tau) : (z,\tau) \in \bar{Q}_i \text{ for some } i \text{ with } \bar{Q}_i \cap K' \neq \emptyset\}$$

and thereupon (3.29) with $K_0 = K''$. We then obtain as in the remark after (3.29) that Lemma 3.33 is true with K replaced by K''. Using the doubling property of ω_2 we get Lemma 3.33 for K. $\square$

4. EXTRAPOLATION

Recall from Lemma 2.1 that for $(Y,s) \in (0,d) \times Q_d(x,t)$

$$L(Y,s) = [\, y_0\,|B|^2 + y_0\,|\nabla A|^2 + y_0^3\,|\tfrac{\partial A}{\partial \tau}|^2$$

$$+ \sum_{i,j=0}^{n-1} \left(\sum_{l=0}^{n-1} y_0\,|\nabla e_l^{ij}|^2 + y_0^{-1}\,|f_l^{ij}|^2 \right) + |g^{ij}|\,]\,(Y,s).$$

Next for (Y,s) as above let

$$D(Y,s) = Q_{y_0/2}(Y,s) \cap [(0,d) \times Q_d(x,t)]$$

$$L_*(Y,s) = L(Y,s) + y_0^{-n-2} \int_{D(Y,s)} L(Z,\tau)dZd\tau$$

$$d\mu_*(Y,s) = L_*(Y,s)dYds.$$

Since $\mu_* \geq \mu^*$, we see that Lemma 2.1 remains valid with μ^* replaced by μ_* in (*iii*). In this section we prove Theorem 1.10. We shall " extrapolate " this theorem from Lemma 2.1 by a bootstrap type procedure. We first prove

Lemma 4.1. *Lemma 2.1 is valid with μ^*, ϵ_2 replaced by μ_*, K whenever $0 < K < \infty$ provided $\eta_i = \eta_i(K, \gamma_1, M, \Lambda, \Lambda_1, n) > 0$ are small enough for $i = 0, 1$.*

Proof: We prove Lemma 4.1 by an induction type argument on K. To avoid confusion we temporarily indicate the dependence of ϵ_2 on the quantities in Lemma 2.1. From this lemma we see that Lemma 4.1 is valid with μ^* replaced by μ_* whenever $K \leq \epsilon_2(\gamma_1, M, \Lambda, \Lambda_1, n)$. Suppose that whenever $\gamma_1, M, \Lambda, \Lambda_1$ are given as above we have shown that Lemma 4.1 holds with μ^* replaced by μ_* whenever $K \leq K^*$ and $K^* \geq \epsilon_2(\gamma_1, M, \Lambda, \Lambda_1, n)$ where $K^* = K^*(\gamma_1, M, \Lambda, \Lambda_1, n)$. We assume as we may that M, Λ, Λ_1 are all ≥ 100. We then put

(4.2)

$$\eta = \left[\frac{\epsilon_2(\frac{1}{2}\gamma_1, 4M, 4\Lambda, 40\Lambda_1, n)}{(\Lambda + \Lambda_1 + M)(1 + K^*)c_1(n)} \right]^{2^{100n}}$$

$$\delta = \left[\frac{\epsilon_2(\frac{1}{2}\gamma_1, 4M, 4\Lambda, 40\Lambda_1, n)}{(\Lambda + \Lambda_1 + M)(1 + K^*)c_1(n)} \right]^{20}$$

and shall show for $c_1 = c_1(n) \geq 1$ large enough that Lemma 4.1 is valid for $K \leq (1 + \eta)K^*$ provided $\eta_i = \eta_i(K)$, $i = 0, 1$ are defined suitably for $K^* < K \leq (1 + \eta)K^*$. We then get Lemma 4.1 by induction. To this end choose N such that $2^{-(N+1)} \leq \delta^5 \leq 2^{-N}$ and recall the definition of K above Lemma 2.1. Put $H \equiv K$ when (a) holds in (iii) of Lemma 2.1 while $H \equiv L_*$ when (b) in (iii) of Lemma 2.1 is valid. Suppose first that

(4.3)
$$\int_{2^{-2N}d}^{d} \left(\int_{Q_d(x,t)} H(Z, \tau) \, dz d\tau \right) dz_0 \geq \eta K^* |Q_d(x,t)|.$$

Using the bisection method we can divide $\bar{Q}_d(x, t)$ into closed rectangles with disjoint interiors and of side length $2^{1-2N}d$, $2^{1-4N}d^2$ in the space and time variables respectively . Let $\{\bar{Q}_j\}$ be these rectangles. From (4.3) and our induction assumption we see that

$$\sum_j \int_0^{2^{-2N}d} \left(\int_{Q_j} H(Z, \tau) \, dz d\tau \right) dz_0 \leq K^* |Q_d(x,t)|$$

which implies by weak type estimates that

$$\int_0^{2^{-2N}d} \left(\int_Q H(Z, \tau) \, dz \, d\tau \right) dz_0 \leq K^* |Q|$$

for some $Q \in \{Q_j\}$. Now we can apply the induction hypothesis with $(0, d) \times Q_d(x, t)$ replaced by $(0, 2^{-2N}d) \times Q$. Using this hypothesis and Harnack's inequality we find that Lemma 4.1 is valid in this case for $K^* < K \leq (1 + \eta)K^*$ provided $\eta_0(K) \leq c_2(n)^{-1} \delta^{10(n+1)} \eta_0(K^*)$, $\eta_1(K) \leq c_3(\gamma_1, \delta, M, n)^{-1} \eta_1(K^*)$ and c_2, c_3 are large enough.

Next suppose that (4.3) is false. We again divide $Q_d(x, t)$ into subrectangles by the bisection method. Let G_m be the closed rectangles obtained in the m th subdivision for $m = 1, 2, \ldots,$. Then the rectangles in G_m have disjoint interiors and side length $2^{1-m}d$, $2^{1-2m}d^2$ in the space and time variables respectively. Let S_m be the subcollection of rectangles $Q_{2^{-m}d}(y, s)$ in G_m with

(4.4)
$$\int_{2^{-(N+j)}d}^{d} \int_{Q_{2^{-j}d}(y,s) \cap Q_d(x,t)} H(Z, \tau) \, dz d\tau dz_0$$

$$\leq (100n)^{100n^2} \eta K^* |Q_{2^{-j}d}(y, s)|$$

$$= \hat{\eta} K^* |Q_{2^{-j}d}(y, s)|$$

for $j = 1, 2, \ldots, m - 1$, while

(4.5)
$$\int_{2^{-(N+m)}d}^{d} \int_{Q_{2^{-m}d}(y,s) \cap Q_d(x,t)} H(Z, \tau) \, dz d\tau dz_0 \geq \hat{\eta} K^* |Q_{2^{-m}d}(y, s)|$$

Using the fact that (4.3) is false and a Calderòn - Zygmund type argument, we get a family of closed rectangles, $S = \bigcup S_m$ with disjoint interiors. Moreover if $(y, s) \notin \bigcup_{Q \in S} Q$, then (4.4) holds for $j = 1, 2, \ldots,$. Put $F^* = Q_d(x, t) \setminus \left(\bigcup_{Q \in S} Q \right)$. We consider two cases : $(a) |F^*| \geq 2\eta |Q_d(x, t)|$ and $(b) |F^*| < 2\eta |Q_d(x, t)|$.

If (a) holds, we suppose $\eta_0(K) \leq \eta/2$ for $K^* < K \leq (1 + \eta) K^*$ and set $d_1 = d[1 - \frac{\eta}{4(n+1)}]$. We observe that there exists F closed, $F \subset F^* \cap E \cap Q_{d_1}(x, t)$ with $|F| \geq \eta |Q_d(x, t)|$. Next we show for each $(y, s) \in F$ and c_1 large enough that

$$(+) \int_{(8\delta)^5 r}^d L(y_0, z, \tau) dy_0 \leq \delta^{100}$$

whenever $(z, \tau) \in Q_r(y, s) \cap Q_d(x, t)$ and $0 < r \leq d$. In fact this inequality follows directly from (4.4), (4.2) if (a) of (iii) in Lemma 2.1 holds. If (b) of (iii) in Lemma 2.1 is valid we use (4.4), (4.2) and weak type estimates to deduce that

$$(++) z_0 L(Z, \tau) \leq \delta^{200}$$

for $(z, \tau) \in Q_r(y, s) \cap Q_d(x, t)$, $(y, s) \in F$ and $(8\delta)^5 r \leq z_0 \leq 3d/4$. Let $\rho = \min\{\delta^{-200} r, 3d/4\}$. We estimate the integral over $[(8\delta)^5 r, \rho]$ in $(+)$ using $(++)$ and the rest of the integral using $(*), (**)$ in Theorem 1.10 (as in (2.3) $(+)$) and the observation that (4.4) implies there exists (z_1, τ_1) with $|z - z_1| + |\tau - \tau_1|^{1/2} \leq 10r$ for which

$$\int_\rho^d L(y_0, z_1, \tau_1) dy_0 \leq \delta^{200}.$$

Let $\hat{\Omega} = \{(Z, \tau) : z_0 > \delta^4 \hat{\sigma}(z, \tau, F)\}$, where $\hat{\sigma}$ is the parabolic distance function defined above (2.2). From $(+), (++)$ we deduce

$$(4.6) \qquad \int_{\delta^4 \hat{\sigma}(z, \tau, F)}^d L(z_0, z, \tau) dz_0 \leq \delta^{100}$$

whenever $(z, \tau) \in Q_d(x, t)$ and

$$(4.7) \qquad\qquad z_0 L(Z, \tau) \leq \delta^{40}$$

for $(Z, \tau) \in \hat{\Omega} \cap [(0, d/2) \times Q_d(x, t)]$. Let $\sigma = \sigma(\cdot, F)$ be the regularized parabolic distance function as in (2.19) and let ξ be as in (2.20). Also let $\theta, 0 \leq \theta \leq 1$, be in $C_0^\infty [(-d/2, d/2) \times Q_{2d}(x, t)]$ with $\theta \equiv 1$ on $(-d/4, d/4) \times Q_d(x, t)$ and

$$d^l \|\tfrac{\partial^l}{\partial x_k^l} \theta\|_\infty + d^{2l} \|\tfrac{\partial^l}{\partial t^l} \theta\|_\infty \leq c(l, n).$$

To handle case (a) we use a slightly more elaborate argument than the one following (2.20). If $(Z, \tau) \in (0, \infty) \times Q_d(x, t)$, put

$$A_1(Z, \tau) = [A(Z, \tau) - A(\delta^3 \sigma(z, \tau, F), z, \tau)] \xi^2(Z, \tau) + A(\delta^3 \sigma(z, \tau), z, \tau),$$
(4.8)
$$B_1(Z, \tau) = B\theta\xi(Z, \tau).$$

Put $A_1 \equiv I, B_1 \equiv 0$ in $U \setminus [(0, \infty) \times Q_d(x, t)]$. Here I denotes the n by n identity matrix. Using (4.6) - (4.8), and (2.19) it can be shown for c_1 sufficiently large, as in the calculations after (2.20), that A_1, B_1 satisfy the hypotheses of Lemma 3.37 of chapter I and Lemma 2.1 with constants $\gamma_1/2, 4M, 4\Lambda, 40\Lambda_1$ (if $(**), (***)$ are valid) and with $Q_d(x, t)$ replaced by $Q_r(y, s)$ whenever $Q_r(y, s) \subset Q_d(x, t)$. Applying Lemma 2.1 we see that the corresponding parabolic measure, ω_1, satisfies the hypotheses of Lemma 3.6. Thus for some $p > 1$, we have $\|\frac{d\omega_1}{dyds}\|_{\alpha_p(Q_d(x,t))} < c < \infty$. Next set

$$\psi = \delta^3 \sigma(\cdot, F),$$

$$\Omega = \{(Z, \tau) : z_0 > \psi\},$$

$$\rho(Z, \tau) = (z_0 + P_{z_0}\psi(z, \tau), z, \tau),$$

when $(Z, \tau) \in U$. Then for c_1 sufficiently large, we see that ρ maps $U, \partial U$ one to one and onto $\Omega, \partial\Omega$. Let v be a solution to the Dirichlet problem for (1.1), A_1, B_1. Then from the remark after Lemma A of chapter I, (2.19) (b), and (2.27) - (2.29) we see that $v \circ \rho$ satisfies (1.1) for some A_2, B_2 satisfying (1.2)-(1.4), as well as the hypotheses of Lemmas 3.14, 3.37 of chapter I. Let ω_1, ω_2 be parabolic measures corresponding to A_1, B_1 and A_2, B_2, respectively. From the above discussion we see that Lemma 3.22 can be applied with $K = F$ to get for some $c' \geq 1$, having the same dependence as η_0 in Lemma 4.1, that

$$c'\, \omega_2(d, x, t + 2d^2, F) \geq 1.$$

We note that ω_2 extends continuously to $U \cup (\partial U \setminus F)$ with $\omega_2 \equiv 0$ on $\partial U \setminus F$. From this remark, the above inequality and the maximum principle for solutions to (1.1) (see the remark after Lemma 3.38 of chapter I) we deduce that if $d_2 = d(1 - \frac{\eta}{8(n+1)})$, then $0 < d_1 < d_2$ and

$$c'\omega_2 \geq 1 \text{ at some point on } (0, r] \times \partial Q_{d_2}(x, t) \cup \{r\} \times Q_{d_2}(x, t)$$

whenever $r > 0$. Next since ω_2 vanishes in $\partial U \setminus \bar{Q}_{d_1}(x, t)$ and $\hat{\sigma}(\partial Q_{d_2}(x, t), Q_{d_1}(x, t)) \approx \eta d$, we can use Lemma 3.9 of chapter I to deduce the existence of $c_+ \geq 1$, having the same dependence as η_0, such that if $0 < r \leq \eta d / c_+$, then $c'\omega_2(\cdot, F) < 1$ on $(0, r) \times \partial Q_{d_2}(x, t)$. Thus for these values of r we have

$$c'\omega_2(r, y, s, F) \geq 1 \text{ for some } (y, s) \in \bar{Q}_{d_2}(x, t).$$

We can now repeat the argument from (2.22) on. Let u_2 be the weak solution to (1.1) corresponding to A_2, B_2 with $u_2 = u \circ \rho$ on ∂U. Then since $u \geq 1$ on F it follows from the definition of ω_2 that for (r, y, s) as above,

$$(4.9) \qquad c'\, u_2(r, y, s) \geq c'\, \omega_2(r, y, s, F) \geq 1.$$

Next we observe that $u \circ \rho, u_2$ satisfy the same pde in $(0, d/4) \times Q_d(x, t)$, since $A_1 = A$, $B_1 = B$ in $\Omega \cap ([0, d/4] \times Q_d(x, t))$. Since $u_2 - u \circ \rho$ vanishes on ∂U we can apply Lemma 3.9 of chapter I to conclude the existence of $c_- > c_+$ such that if $r' = \eta d / c_-$, then

$$(4.10) \qquad c'|u_2(r', y, s) - u \circ \rho(r', y, s)| \leq \tfrac{1}{2}$$

for all $(y, s) \in \bar{Q}_{d_2}(x, t)$. Combining (4.9), (4.10) we conclude first that $c'(u \circ \rho)(r', y, s) \geq \tfrac{1}{2}$ for some $(y, s) \in \bar{Q}_{d_2}(x, t)$ and second from Harnack's inequality that Lemma 4.1 is valid when $K^* \leq K \leq (1 + \eta)K^*$.

Next we consider case (b). We claim there exists a finite subcollection S' of S such that if $Q \in S'$ and $\eta' = \frac{\hat{\eta}}{(100n)^{10n}}$ then

$$(a) \qquad \int_0^{2^{-(N+1)} s(Q)} \int_Q H(Z,s)\, dz d\tau dz_0 \leq (1-\eta')\, K^* \,|Q|,$$

(4.11)

$$(b) \qquad \sum_{Q \in S'} |Q| \geq \eta' \,|Q_d(x,t)|,$$

$$(c) \qquad \hat{\sigma}(Q,Q') \geq 4n \,\max\{s(Q),\, s(Q')\}\,.$$

In $(4.11)(a)$, $s(Q)$ denotes the side length of Q in the space direction. To prove our claim we let

$$(4.12) \qquad\qquad T(Q) = (2^{-(N+1)}\, s(Q),\, d\,) \times Q$$

whenever $Q \in S$ and observe from (4.12) as well as the definition of S that

$$(4.13) \qquad\qquad \int\int_{T(Q)} H\, dZ\, d\tau \geq \hat{\eta}\, K^*\, |Q|.$$

Clearly

$$(4.14) \qquad\qquad T(Q), T(Q'), Q, Q' \text{ have disjoint interiors}$$

whenever $Q, Q' \in S$ and $Q \neq Q'$. From the induction hypothesis and (4.14) we note that

$$\sum_{Q \in S} \int_0^{2^{-(N+1)} s(Q)} \int_Q H(Z,s)\, dz d\tau dz_0 + \sum_{Q \in S} \int\int_{T(Q)} H(Z,s)\, dZ\, ds$$

(4.15)

$$\leq (1+\eta)\, K^*\, |Q|\,.$$

Using (4.12) - (4.15) and the definition of $\hat{\eta}$ we obtain

$$\sum_{Q \in S} \int_0^{2^{-(N+1)} s(Q)} \int_Q H(Z,s)\, dz d\tau\, dz_0 \leq K^*\, (1 - (100n)^{5n}\, \eta')\, |Q_d(x,t)|\,.$$

Now

$$\sum_{Q \in S} |Q| \geq (1-2\eta)|Q_d(x,t)|$$

since $|F^*| < 2\eta$. Claim (4.11) follows from these two inequalities, weak type estimates, and a covering argument.

Next we suppose $\eta_0(K) \leq \eta_0(K^*)\eta^{n+2}$ for $K^* < K \leq (1+\eta)K^*$. We shall show the existence of a finite subset $\hat{S} = \{\, Q_{r_i}(y_i, s_i)\,\}_1^l$ of S' and $Q_{4r_i'}(z_i, \tau_i) \subset$

$Q_{r_i}(y_i, s_i)$ such that for $1 \leq i \leq l$, we have $\frac{1}{2}\eta \leq \frac{r_i'}{r_i} \leq \eta$ and

$$(i) \qquad \int_{(0,r_i') \times Q_{r_i'}(z_i, \tau_i)} H(Z, \tau)\, dZ d\tau \; \leq \; K^* |Q_{r_i'}(z_i, \tau_i)|,$$

$$(ii) \qquad |E \cap Q_{r_i'}(z_i, \tau_i)| \; \geq \; (1 - \eta_0(K^*)) |Q_{r_i'}(z_i, \tau_i)|,$$

(4.16)

$$(iii) \qquad \sum_{i=1}^{l} |Q_{r_i'}(z_i, \tau_i)| \; \geq \; \eta^{n+2} |Q_d(x, t)|,$$

$$(iv) \qquad \text{Either } r \geq \eta d/100n \text{ for some } Q_r(y, s) \in \hat{S} \text{ or}$$
$$\cup_{Q \in \hat{S}} Q \subset Q_{d_1}(x, t).$$

In (iv), d_1 is defined as in case (a). (4.16) (i) is a consequence of $(4.11)(a)$ and our usual weak type argument. (4.16) $(ii), (iii), (iv)$ follow from $(4.11)(b)$ and a counting argument using the definition of $\eta_0(K)$. We omit the details.

First suppose there exists $Q_{r_i}(y_i, s_i) \in \hat{S}$ with $r_i \geq \eta d/100n$. Then from (4.16) $(i), (ii)$ and the induction hypothesis we see that $c\, u(r_i', z_i, \tau_i + 2(r_i')^2) \geq 1$ for some c having the same dependence as η_0. From this inequality and Harnack's inequality we conclude that Lemma 4.1 is valid for $K^* < K \leq (1 + \eta)K^*$. Thus we assume that the second alternative in (4.16) (iv) occurs.

Put $F_+ = \{(y_i, s_i) : Q_{r_i}(y_i, s_i) \in \hat{S}\}$ and let

$$\tilde{\sigma}(z, \tau) = \begin{cases} r_i \text{ when } |z - y_i| + |\tau - s_i|^{1/2} \leq r_i, \ 1 \leq i \leq l, \\ = \hat{\sigma}(z, \tau, F_+), \text{ otherwise in } R^n. \end{cases}$$

We now are in a position to essentially repeat the argument in case (a) from (4.6) on. More specifically we define $\hat{\Omega}$ as in case (a) relative to $\tilde{\sigma}$ and use (4.4), to show that (4.6), (4.7) are valid. Let $\sigma_+ \in C_0^\infty(R^n)$ be a parabolic regularization of $\tilde{\sigma}$ constructed so that (2.19) $(a)-(c)$ are valid with $\sigma, \hat{\sigma}$ replaced by $\sigma_+, \tilde{\sigma}$, respectively. Define A_1, B_1 as in (4.8) with σ replaced by σ_+. Then as in the first case we see from (4.6), (4.7), for c_1 large enough, that A_1, B_1 satisfy the hypotheses of Lemma 2.1 with $Q_d(x, t)$ replaced by $Q_r(y, s)$ whenever $Q_r(y, s) \subset Q_d(x, t)$. Consequently, Lemma 3.37 of chapter I, (3.1), and Lemma 3.6 are valid for ω_1. Next we define ψ, ρ, Ω relative to σ_+ as in case (a). Let v be a solution to (1.1) corresponding to A_1, B_1. Then from (2.19) (b), (2.25)-(2.28), and the remark after Lemma A we see that $v \circ \rho$ satisfies (1.1) relative to some A_2, B_2 and Lemma 3.37 is valid for the corresponding parabolic measure, ω_2. Next from $(4.11)(c)$ and the definition of $\tilde{\sigma}$ we see that Lemma 3.33 can be applied with $K = \bigcup_{Q \in \hat{S}} \bar{Q}$. Applying this lemma we get for some $c'' \geq 1$ having the same dependence as η_0,

$$c''\omega_2(d, x, t + 2d^2, \cup_{Q \in \hat{S}} \bar{Q}) \; \geq \; 1.$$

Now as above we see from (4.16) and the induction hypothesis that for each i we have $c\, u(r_i', z_i, \tau_i + 2(r_i')^2) \geq 1$. Using this fact and Harnack's inequality we conclude that there exists $c^* \geq 1$, for which $c^* u \circ \rho \geq 1$ on $Q_{r_i'}(z_i, \tau_i + 4(r_i')^2)$, $1 \leq i \leq l$. Let u_2 be the solution to the continuous Dirichlet problem for (1.1), A_2, B_2, with $u_2 \equiv u \circ \rho$ on ∂U. From the previous inequality for $u \circ \rho$ we deduce that

$$(4.17) \qquad c^* u_2 \; \geq \; \omega_2(\cdot, \cup Q_{r_i'}(z_i, \tau_i + 4(r_i')^2)).$$

Also from the above inequality for ω_2 and Lemma 3.37 (β) we find for some $c^{**} \geq 1$ with the same dependence as η_0 that

$$(4.18) \qquad c^{**}\,\omega_2(d, x, t + 2d^2, \cup Q_{r_i'}(z_i, \tau_i + 4(r_i')^2)) \geq 1.$$

Using (4.17), (4.18), we can now argue as in case (a) to get first (4.9) with F replaced by $\bigcup Q_{r_i'}(z_i, \tau_i + 4(r_i')^2)$ and then (4.10). As in case (a) we conclude from (4.9), (4.10) that Lemma 4.1 is true when $K^* < K \leq (1 + \eta)K^*$. We put $\eta_0(K) = \eta_0(K^*)\eta^{n+2}$ and observe for this value of η_0, that Lemma 4.1 is true for the above values of K. By induction we now obtain Lemma 4.1. $\square$

To finish the proof of Theorem 1.10 we need to show the continuous Dirichlet problem corresponding to A, B always has a unique solution. Indeed let $B_j(X, t) = B(x_0 + j^{-1}, x, t)$ for $j = 1, 2, \ldots$, and $(X, t) \in U$. Now B_j converges pointwise to B almost everywhere as $j \to \infty$ and A, B_j satisfy the hypotheses of Lemma 4.1 whenever $Q_d(x, t) \subset R^n$ with constants independent of j since $c\mu^*[(0, d) \times Q_d(x, t)] \geq \mu_*[(0, d) \times Q_d(x, t)]$ for some $c = c(n)$ as follows from interchanging the order of integration in the integral defining μ_*. Also since B_j is essentially bounded we can use the remark after (3.22) of chapter I and the same argument as in $(i) - (iv)$ of Lemma 3.37 in chapter I to deduce that the continuous Dirichlet problem for A, B_j always has a unique solution. From this fact and Lemma 4.1 we see that (3.1) holds for the corresponding parabolic measures with constants that are independent of j. Thus Lemma 3.2 is valid with A_1, B_1 replaced by A, B_j and with constants that are independent of j. Lemma 3.2 with uniform constants can be used to show that the continuous Dirichlet problem corresponding to A, B has a unique solution (this is the gist of $(i) - (iv)$ of Lemma 3.37). We can now use Lemma 4.1 for A, B to conclude first that (3.1) is true for the corresponding parabolic measure and second from Lemma 3.6 that Theorem 1.10 is true. $\square$

Proof of Theorem 1.13. Finally in this section we show that Theorem 1.10 implies Theorem 1.13. We again prove a more general result for use in chapter III.

Lemma 4.19. *Let A_1, B_1, ω_1 be as in the hypotheses of Lemma 3.6 and let $p > 1$ be as in the conclusion of this lemma. If $q = p/(p - 1)$, then the $L^q(R^n)$ Dirichlet problem for A_1, B_1 always has a unique solution in the sense of (I) and (II) of Theorem 1.13.*

Proof: To prove this lemma we note first from Harnack's inequality that whenever $k \geq 1$ and $E \subset R^n$ is a Borel set, then

$$\omega_1(d, x, t + 2d^2, E) \leq c(k, \gamma, M, n)\,\omega_1(kd, x, t + 2k^2d^2, E).$$

Since $\frac{d\omega_1}{dyds}(kd, x, t + 2k^2d^2, \cdot) \in \alpha_p(Q_{kd}(x, t))$ with norm $\leq c$ for $k = 1, 2, \ldots$, we see that $\omega_1(d, x, t + 2d^2, \cdot)$ is absolutely continuous with respect to Lebesgue measure on R^n whenever $d > 0$ and $(x, t) \in R^n$. Thus if we put

$$\frac{d\omega_1}{dyds}(d, x, t + 2d^2, \cdot) = K(d, x, t + 2d^2, \cdot),$$

then

$$(4.20) \qquad \int_{Q_{kd}(x,t)} K^p(2kd, x, t + 4k^2d^2, y, s)\,dyds \leq c\,|Q_{kd}(x, t)|^{1-p},$$

for some $c = c(\gamma_1, M, \eta_0, \eta_1, n)$ and $k = 1, 2, \ldots,$. For our estimates we shall use the fact that (4.20) actually holds with p replaced by some $p_1 > p$ and c suitably large(see[Gi, ch 5]). We suppose first that $f \in C_0^\infty(R^n) \cap L^q(R^n)$ and set

$$u(X,t) = \int_{\mathbb{R}^n} K(X,t,y,s)\, f(y,s)\, dy\, ds,$$

whenever $(X,t) \in U$. Clearly u is a weak solution to (1.1) in U. Now the continuous Dirichlet problem for (1.1), A_1, B_1 has a solution corresponding to f and this solution is unique thanks to the maximum principle in Lemma 3.38 of chapter I. Using basic functional analysis arguments we see that u is this solution so u extends continuously to $\bar{U}$ and $u \equiv f$ on ∂U. We prove (II) only for Nf in (2.14) of chapter I defined relative to parabolic cones with $a = 1$. The proof for general $a > 0$ is similar. We assume as we may that $f \geq 0$. Given $(x,t) \in R^n$ and $(Y,s) \in \Gamma_1(x,t)$, we let $d = y_0$. Next we choose a sequence of continuous functions on R^n with

$$(a) \qquad \phi_0 \equiv 1 \text{ on } Q_{2d}(x,t) \text{ and supp } \phi_0 \subset Q_{4d}(x,t),$$

$$(b) \qquad \phi_j \equiv 1 \text{ on } Q_{2^{j+1}d}(x,t) \setminus Q_{2^j d}(x,t) \text{ for } j = 1, \ldots,$$
$$\text{and supp } \phi_j \subset Q_{2^{j+2}d}(x,t) \setminus Q_{2^{j-1}d}(x,t) \text{ for } j = 1, \ldots,$$

$$(c) \qquad 0 \leq \phi_j \leq 1, \text{ for } j = 1, \ldots, .$$

From Harnack's inequality we observe that

$$c^{-1}u(Y,s) \leq u(d,x,t+2d^2) \leq \sum_{j=0}^{\infty} \int_{\mathbb{R}^n} f\phi_j \, dz d\tau$$

(4.21)

$$= \sum_{j=0}^{\infty} u_j(d,x,t+2d^2),$$

where c has the same dependence as the constant in (4.20). (4.21) is essentially the same as (5.9) of chapter I. We can in fact repeat the argument after (5.9) to deduce first from Lemma 3.2 that

$$u_j(d,x,t+2d^2) \leq c2^{-j\beta}u_j(2^{j+3}d,x,t+2^{2j+6}d^2),$$

and second from (4.20) for $p_1 > p$ that

$$u_j(2^{j+3}d,x,t+2^{2j+6}d^2)^{q_1} \leq cM(f^{q_1})(x,t)$$

where $q_1 = \frac{p_1}{p_1-1}$. Using the above inequalities in (4.21) and the Hardy - Littlewood maximal theorem we see that (II) of Theorem 1.13 is valid when $f \in C_0^\infty(R^n) \cap L^q(R^n)$. The general case $f \in L^q(R^n)$ follows from the smooth case and the basic estimates in Lemmas 3.3,3.4 of chapter I. (I) is easily deduced from (II) and the fact that (I) is valid when f is continuous. Since this argument is well known we omit the details.

   To prove uniqueness, suppose u, v are both solutions to (1.1) satisfying $(I), (II)$ relative to $f \in L^p(R^n)$. Then from (1.4) and our knowledge of parabolic pde's with constant coefficients we see there exists $r > 0$ such that $u - v$ has a bounded continuous extension to $\bar{U} \setminus Q_r(0,0)$ with $u - v \equiv 0$ on $\partial U \setminus Q_r(0,0)$. Next given $\epsilon > 0$ and $(d,x,t+2d^2) \in U$ we see from this fact and a measure theoretic argument (essentially Ergoff's theorem) that there exists a compact set $K = K(\epsilon) \subset Q_r(0,0) \subset R^n$ such that if $\psi = \hat{\sigma}(\cdot, K \cup [R^n \setminus Q_r(0,0)])$ and

$\rho(Y,s) = (y_0 + P_{\gamma y_0}\psi(y,s), y, s), (Y,s) \in \bar{U},$ then

(4.22)

(a) $|Q_r(0,0) \setminus K| \leq \epsilon,$

(b) $(u-v) \circ \rho$ is continuous in $\bar{U}$ with $(u-v) \circ \rho = 0$ on $K \cup (R^n \setminus Q_r(0,0)).$

We note that $\sigma \leq c(n)\epsilon^{1/(n+1)}$. Consequently for ϵ small enough it follows from the remark after Lemma A and (2.27)-(2.29) that $(u-v) \circ \rho$ satisfies weakly in U a pde of the form (1.1) corresponding to some A_2, B_2 for which (1.2)-(1.4) are valid. We also assume that $c(n)\epsilon^{1/(n+1)} \leq d/2$ so that $(d/2, x, t+2d^2) \in \rho(U)$. The idea now is to construct $w \geq 0$ a solution to the Dirichlet problem for (1.1) corresponding to A_1, B_1 with

(*) $|(u-v) \circ \rho| \leq w \circ \rho$ on $\partial U,$

(4.23)

(**)   For some fixed $a > 0$, $\|N_a w\|_{L^q(\mathbb{R}^n)} = \xi$ where $\xi \to 0$ as $\epsilon \to 0.$

Here N_a stands for the nontangential maximal function defined relative to a as in (2.14) of chapter I. From (*) and the maximum principle in Lemma 3.38 of chapter I we see that $(u-v) \circ \rho \leq w \circ \rho$ on U. Using this fact and Harnack's inequality we deduce from (**) that

$$|u-v|(d,x,t+2d^2)^q \leq w(d,x,t+2d^2)^q$$

$$\leq c\,d^{-(n+1)} \int_{Q_d(x,t+4d^2)} w^q(d,z,\tau)\,dz d\tau$$

$$\leq c\,d^{-(n+1)} \|N_a w\|^q_{L^q(\mathbb{R}^n)} \leq c\,d^{-(n+1)} \xi^q,$$

where the c's depend only on γ_1, M, n. Since $\epsilon > 0$ and $(d,x,t+2d^2)$ are arbitrary we conclude that $u \equiv v$. Thus uniqueness holds in Lemma 4.19 once we have proved (4.23).

   To construct w we use a more or less standard argument. Given $\delta > 0$, let O be an open set with $\bar{O} \subset Q_r(0,0) \setminus K$ such that $|(u-v) \circ \rho| \leq \delta$ on $R^n \setminus O$. Existence of O follows from (4.22). Let $\{Q_i\}$ be a Whitney decomposition of $Q_r(0,0) \setminus K$ into parabolic rectangles with disjoint interiors and side lengths proportional to their distance from $K \cup \partial Q_r(0,0)$. Let $h \geq 0$ be a continuous function on R^n defined as follows. If $Q = Q_{\hat{r}}(\hat{y},\hat{s}) \in \{Q_i\}$ and $Q \cap O \neq \emptyset$ let $h \equiv \inf\{\delta + N_a(u-v)(z,\tau) : (z,\tau) \in Q_{\hat{r}/4}(\hat{y}, \hat{s} - \hat{r}^2/4)\} = b$ on $Q_{\hat{r}/4}(\hat{y}, \hat{s} - \hat{r}^2/4)$. Extend h continuously to the rest of $\bar{Q}$ in such a way that $h \leq b$ and $h \equiv \delta$ on $Q \setminus Q_{\hat{r}/4}(\hat{y}, \hat{s} - \hat{r}^2/4)$. If $Q \cap O = \emptyset$ we set $h \equiv \delta$ on Q. We also put $h \equiv \delta$ on K and $Q_{2r}(0,0) \setminus Q_r(0,0)$. Finally we extend h continuously to the rest of R^n in such a way that h is nonnegative with supp $h \subset Q_{4r}(0,0)$. Let

$$w(Y,s) = c' \int_{\mathbb{R}^n} h(z,\tau) K(Y,s,z,\tau)\,dz d\tau$$

when $(Y,s) \in U$. We reserve our choice of a and c' until later. Then w is the solution to the continuous Dirichlet problem for (1.1) corresponding to A_1, B_1 with $w = c' h$ on ∂U. To prove (*) we observe first from positivity of h that

$$0 = |(u-v) \circ \rho| \leq w \circ \rho \text{ on } K \cup (R^n \setminus Q_r(0,0)).$$

If $Q_{\hat{r}}(\hat{y}, \hat{s}) \in \{Q_i\}$, then there exists $Q_{r'}(y', s') \in \{Q_i\}$ with $s' \leq \hat{s} - \hat{r}^2$ and $\bar{Q}_{r'}(y', s') \cap \bar{Q}_{\hat{r}}(\hat{y}, \hat{s}) \neq \emptyset$. From the definition of h, w we find for $(y, s) \in \bar{Q}_{\hat{r}}(\hat{y}, \hat{s})$ and a, c' large enough (depending only on $\gamma_1, M, \eta_0, \eta_1, n$) that

$$
w \circ \rho(y, s) \;\geq\; c' \int_{Q_{r'}(y', s')} K(\rho(y, s), z, \tau) h(z, \tau) \, dz d\tau
$$

$$
\geq \; c' \, |(u - v) \circ \rho|(y, s) \int_{Q_{r'/4}(y', s' - (r')^2/4)} K(\rho(y, s), z, \tau) \, dz d\tau
$$

$$
\geq \; |(u - v) \circ \rho|(y, s).
$$

Here we have used (3.1) with $E = Q_{r'/4}(y', s' - (r')^2/4)$ and Harnack's inequality. Thus $(*)$ is true. $(**)$ follows from (II) and properties of the Lebesgue integral. The proof of Lemma 4.19 is now complete. $\square$

To prove Theorem 1.13 we simply observe from Theorem 1.10 and Lemma 4.1 that we may replace A_1, B_1 by A, B in the hypotheses of Lemma 4.19. $\square$

Remark. It would be interesting to know if Lemma 4.1 has a simpler proof when $E = Q_d(x, t)$. This special case of Lemma 4.1 is all that was needed to prove Hölder continuity for solutions vanishing continuously on $Q_d(x, t)$ in Lemma 3.2. Perhaps Lemma 3.2 is even valid with less restrictive assumptions on A. More specificallly is the conclusion of Lemma 3.2 valid if we assume only that A, B satisfy (1.2)-(1.4) and B satisfies (1.5)?

As regards the doubling property of ω mentioned in the remark after the statement of Theorem 1.10, we note that if $\nabla \cdot B = 0$ weakly, then the adjoint pde (3.1) is similar to (1.1), so for example Lemma 3.2 holds with (1.1) replaced by (3.1). In this case we have managed to use the method in [FS] to prove that parabolic measure corresponding to (1.1) is doubling. We omit the details of the proof in this memoir. For further remarks concerning parabolic doubling see the remarks after sections 10 and 11 of this memoir.

5. References

[CF] R. Coifman and C. Fefferman, *Weighted norm inequalities for maximal functions and singular integrals*, Studia Math. **51** (1974), 241-250.

[CFMS] L. Caffarelli, E. Fabes, S. Mortola, and S. Salsa, *Boundary behavior of nonnegative solutions of elliptic operators in divergence form*, Indiana Univ Math. J. **30** (1981), 621-640.

[DJK] B. Dahlberg, D. Jerison, and C. Kenig, *Area integral estimates for elliptic differential operators with nonsmooth coefficients*, Ark. Mat. **22** (1984), 97-108.

[FS] E. Fabes and M. Safonov, *Behaviour near the boundary of positive solutions of second order parabolic equations*, J. Fourier Anal. Appl. **3** (1997), 871-882.

[G] F. Gehring, L^p *integrability of the partial derivatives of a quasiconformal mapping*, Acta Math. **130** (1973), 265-277.

[Gi] M. Giaquinta, *Multiple integrals in the calculus of variations and nonlinear elliptic systems*, Annals of Mathematics Studies **105**, Princeton University Press. 1983.

[H] Y. Heurteaux, *Solutions positives et mesure harmonique pour des opérateurs paraboliques dans des ouverts " lipshitziens, "* Ann. Inst. Fourier (Grenoble) **41** (1991), 601-649.

[KKPT] C.Kenig, H. Koch, J. Pipher, and T. Toro, *A new approach to absolute continuity of elliptic measure,with applications to nonsymmetric equations*, submitted.

[LM] J. Lewis and M. A. Murray, *The method of layer potentials for the heat equation in time-varying domains*, Memoirs of the AMS. **545** (1995), 1-157.

[St] E. Stein, *Singular integrals and differentiability properties of functions*, Princeton University Press, Princeton, 1970.

CHAPTER III
ABSOLUTE CONTINUITY AND THE L^p DIRICHLET PROBLEM: PART 2

1. INTRODUCTION

In this chapter we consider parabolic generalizations of a theorem of [FKP]. To this end, recall that throughout this memoir we have considered weak solutions u to pde's of the form

$$(1.1) \qquad Lu = u_t - \nabla \cdot (A\nabla u) - B\nabla u = 0$$

in U under the following structure assumptions on A, B.

$$(1.2) \qquad \langle A(X,t)\xi, \xi \rangle \geq \gamma_1 |\xi|^2$$

for some $\gamma_1 > 0$, almost every $(X,t) \in U$ and all $n \times 1$ matrices ξ.

$$(1.3) \qquad \left(\sum_{i=0}^{n-1} x_0 |B_i| + \sum_{i,j=0}^{n-1} |A_{ij}| \right)(X,t) < M < \infty$$

for almost every $(X,t) \in U$. For some large $\rho > 0$,

$$(1.4) \qquad A \equiv \text{ constant matrix in } U \setminus Q_\rho(0,0).$$

We also assume for given A, B satisfying (1.2)-(1.4) and some λ, p, $1 < \lambda, p < \infty$ that

(1.5)

  (a) The continuous Dirichlet problem for (1.1), A, B has a unique solution,

  (b) If ω is parabolic measure for (1.1), A, B, then $\lambda\omega(d, x, t + 2d^2, Q_d(x,t)) \geq 1$,

  (c) $\| \frac{d\omega}{dyds}(d, x, t + 2d^2, \cdot) \|_{\alpha_p(Q_d(x,t))} < \lambda < \infty$,

for all $(x,t) \in R^n$, $d > 0$. Next given $(X,t) \in U$ and A_1, A_2, B_1, B_2 satisfying (1.2)-(1.4) put

$$d\nu(X,t) = \text{ ess sup } \left\{ x_0^{-1} |A_1 - A_2|^2(Y,s) + x_0 (|B_1|^2 + |B_2|^2)(Y,s) \right.$$

$$\left. : (Y,s) \in Q_{x_0/2}(X,t) \right\} dXdt.$$

We assume that ν is a Carleson measure on U with

$$(1.6) \qquad \|\nu\| \leq \hat{\beta} < \infty.$$

With this notation we prove in sections 1 and 2, the following parabolic analogue of Theorem 2.3 in [FKP].

Theorem 1.7. *Let A_1, A_2, B_1, B_2, satisfy (1.2)-(1.4) and (1.6). If (1.5) holds for A_1, B_1 and some λ_1, p_1, then there exists $\lambda_2, p_2 > 1$ such that (1.5) is valid with A, B, λ, p replaced by A_2, B_2, λ_2, p_2. Moreover, if $q_2 = p_2/(p_2 - 1)$, then for*

$q_2 \leq q < \infty$, *the $L^q(\mathbf{R}^n)$ Dirichlet problem for (1.1), A_2, B_2, always has a unique solution in the sense of $(I), (II)$ of Theorem 1.13 in chapter 2.*

Remark 1) We note that Nyström [N] has obtained analogues of theorems in [FKP] for pde's of the form (1.1) when $B \equiv 0$ in Lip (1, 1/2) domains. He does this by first showing that the argument in [FS] (see chapter I) can be generalized to Lip (1, 1/2) domains in order to obtain that the corresponding parabolic measures are doubling. After getting doubling, he is able to use a proof modeled on the one of [FKP]. Our situation is quite different as we have not been able to prove that (1.5) implies ω is doubling, which is an important ingredient in the proof of [FKP]. Again the main reason we cannot prove doubling (as in Theorem 1.10 of chapter II) is because we cannot prove basic estimates for the adjoint pde corresponding to (1.1). This lack of doubling considerably complicates our proofs. For example we were not able to get an L^r bound for the area function in terms of the nontangential maximal function or vice versa (as concerns solutions to (1.1) corresponding to A_1, B_1) and thus were forced to devise a proof different from [FKP] which does not use these bounds.

2) As for a proof we follow the general strategy of Theorem 1.10 of chapter II and first show that this theorem is valid when a certain measure involving $A_1 - A_2$ and $B_1 - B_2$ satisfies a Carleson measure condition similar to the one considered in [Fe] and [FKP], with small Carleson norm. In order to do this we begin by proving existence and making some basic estimates for the Green's function corresponding to A_1, B_1 (see Lemma 2.2). We then study solutions to the equation $Lu = \nabla \cdot F + f$ in Lemmas 2.6, 2.10. Finally we use these lemmas, as well as Picard iteration, to first prove Lemma 2.23 and second in Lemma 2.42 that Theorem 1.7 is valid in this special case.

3) After proving the above special case we consider the general case of Theorem 1.7. Our argument in this case is necessarily more complicated than the one in [FKP]. To see why we note that these authors get the large Carleson norm special case considered in [Fe] (see Theorem 2.4 in [FKP]) simply by applying their estimates in the small norm case to $t \to tA_2 + (1-t)A_1$, $0 < t < 1$, on short intervals depending on the Carleson norm. This argument works because the above authors have basic estimates for their pde's which involve constants depending only on the ellipticity constants for A_1, A_2. Unfortunately our basic estimates depend in addition to the ellipticity constants on λ in $(1.5)(b)$. Thus our estimates can vary with t and so could conceivably blow up for some $t < 1$.

To overcome this difficulty we extract the case when either $A_1 \equiv A_2$, $B_2 \equiv 0$ or $B_1 \equiv 0$, from the special case mentioned in 1), by an induction type argument similar to the one we use in proving Theorem 1.10 of chapter II. Again the comparison lemmas for parabolic measure in section 3 of chapter II will play an important role in the proof (see the remark after the statement of Theorem 1.10). The above two cases are easily seen to imply Theorem 1.7.

4) Theorem 1.7 is proved in sections 2 and 3. For possible generalizations of this theorem see the remark at the end of section 3.

5) For elliptic operators we can prove a stronger version of Theorem 1.7. In order to state this theorem let A_1, A_2, B_1, B_2 satisfy $(1.2) - (1.4)$ with (X, t) replaced by X and suppose that (1.5) is also valid with $(x, t + 2d^2), Q_d(x, t)$, replaced by $x, B_d(x)$

and ω = elliptic measure corresponding to A_1, B_1. Set

$$d\tilde{\nu}(X) = \text{ess sup } \left\{ x_0^{-1} |A_1 - A_2|^2(Y) + x_0 (|B_1 - B_2|^2)(Y) \right.$$

$$\left. : Y \in B_{x_0/2}(X) \right\} dX dt.$$

We assume that $\tilde{\nu}$ is a Carleson measure on $\hat{U}$ with

(1.8) $$\|\tilde{\nu}\| \leq \tilde{\beta} < \infty.$$

With this notation we prove in section 4, the following analogue of Theorem 2.3 in [FKP].

Theorem 1.9. *Let A_1, A_2, B_1, B_2, be as above. Then there exists $\lambda_2, p_2 > 1$ such that (1.5) is valid with $A, B, \lambda, p, (x,t), \alpha_p(Q_d(x,t))$ replaced by $A_2, B_2, \lambda_2, p_2, x, \beta_p^*(B_d(x))$. Moreover, if $q_2 = p_2/(p_2 - 1)$, then for $q_2 \leq q < \infty$, the $L^q(R^{n-1})$ Dirichlet problem always has a unique solution in the sense of $(I), (II)$ of Theorem 1.15 of chapter II.*

Remark. 1) Theorem 1.9 when $B_1 \equiv B_2 \equiv 0$ is equivalent to Theorem 2.3 in [FKP]. Our proof of Theorem 1.9 is different than [FKP] even in the case when $B_1 \equiv B_2 \equiv 0$. As mentioned in remark 1) after Theorem 1.10, our proof avoids the use of S and N functions. Also when B_1 or $B_2 \not\equiv 0$, we are forced to give a more complicated argument in the large Carleson norm case of Theorem 1.9 (see remark 3) after Theorem 1.7). The reader is invited to compare the two arguments.
2) Note that (1.8) is a weaker assumption than (1.6). Again we can prove a stronger theorem in the elliptic case mainly because in Lemma 4.6 we shall show that an elliptic measure satisfying (1.8) is necessarily doubling. Our proof of doubling differs from the usual proof where one estimates $\omega(d, x, B_r(y))$ in terms of $r^{2-n}G(d, x, r, y)$ when $B_{2r}(y) \subset B_d(x)$, by choosing an appropriate test function and using essentially subsolution estimates in $(-r, r) \times B_r(y)$. This proof is not available in our situation so instead we use an iterative procedure to obtain the above estimate. Finally in section 4 we prove Theorems 1.14-1.15 stated in chapter II. For a closing remark on parabolic doubling see the remark at the end of section 4.

The authors would like to thank Carlos Kenig and Jill Pipher for sharing their work with them at an early stage (see the remark following the statement of Theorem 1.14 in chapter II).

2. PROOF OF THEOREM 1.7 IN A SPECIAL CASE

As mentioned in section 1, the proof of Theorem 1.7 will proceed in the same way as the proof of Theorem 1.10. That is we first prove this theorem in a special case (Lemma 2.42) and then in section 3 extrapolate the general case from this special case using Lemmas 3.6, 3.22, and 3.33 of chapter II. On the one hand Lemma 2.42 is more difficult to prove than Theorem 2.13 in chapter I in the sense that we do not know to begin with whether it suffices to consider only smooth $A_1, B_1.A_2, B_2$. Thus we must review our basic estimates in section 3 of chapter I and establish the existence of the Green's function as well as some of its properties for (1.1) corresponding to A_1, B_1. On the other hand once we do our preliminary investigations, most of the remainder of the proof (see Lemma 2.23) will involve estimating the terms in a certain iterative sequence wheareas the proof of Theorem 2.13 involved numerous integrations by parts. We note as in section 1 that our proof manages to avoid any $L^p(\partial U)$ estimates of Su in terms of Nu. In fact we have not been able to determine whether such estimates are valid for solutions u to (1.1) (corresponding to A_1, B_1) with $L^p(\partial U)$ nontangential limits. The essential difficulty in trying to prove such estimates as in [DKJ], is that we do not know whether parabolic measure corresponding to A_1, B_1 restricted to a parabolic sawtooth domain is doubling, so that we cannot apply Lemma 3.22 of chapter II. Luckily for our extrapolation scheme to work we do not need such estimates.

To begin let A, B satisfy (1.2)-(1.4) and suppose that the continuous Dirichlet problem for (1.1), A, B, always has a unique solution. If ω denotes the corresponding parabolic measure we assume for some positive $c^* < \infty$ that

$$(2.1) \qquad\qquad c^*\, \omega(d, x, t + 2d^2, Q_d(x, t)) \geq 1$$

whenever $d > 0$ and $(x, t) \in R^n$. Our first lemma is

Lemma 2.2. *Let A, B be as above and suppose that ω satisfies (2.1). There exists $G : U \times U \to R$ with the following properties. If $(X, t), (Y, s) \in U$, $(X, t) \neq (Y, s)$, and $r = |X - Y| + |s - t|^{1/2}$, then for some $c \geq 1$, $0 < \theta < \frac{1}{2}$ (depending only on γ_1, M, n, c^*), we have*

$(a) \qquad G(X, t, Y, s) \leq c r^{-n}, 0 < r < y_0/2,$

$(b) \qquad G(X, t, Y, s) \leq c\, y_0^{-n}\, \omega(X, t, Q_{r_1}(z, \tau))$ *for* $(z, \tau) \in Q_{y_0/16}(y, s)$, $r > y_0/2, y_0/4 < r_1 < y_0,$

$(c) \qquad G(X, t, Y, s) < c\, (x_0/r)^\theta\, G(\hat{X}, \hat{t}, Y, s)$ *for* $r > x_0$ *and* $\hat{X} = (x_0 + r, 0, \dots, 0), \hat{t} = t + 2r^2,$

$(d) \qquad G(\cdot, Y, s)$ *and* $G(Y, s, \cdot)$ *are weak solutions to*(1.1), (3.1) *respectively (see chapter I) in* $U \setminus \{(Y, s)\},$

$(e) \qquad$ *If* $0 < d_1, d_2 < \min\{r/100n, y_0/2, x_0/2\}$, *then* $G(\cdot, \cdot)$ *is Hölder continuous on* $Q_{d_1}(X, t) \times Q_{d_2}(Y, s)$, *with exponent independent of* $d_1, d_2, (X, t), (Y, s).$

Proof: Recall that in section 3 of chapter I (see (3.6), (3.7)) we defined the Green's function $\hat{G}$ relative to $\hat{A}, \hat{B}$ satisfying (1.2)-(1.4) when either $\hat{B} \equiv 0$ or $\hat{A}, \hat{B}$ are smooth. Moreover $(d), (e)$ were valid with G replaced by $\hat{G}$. Also for $(Y, s) \in U$ the functions $\hat{G}(Y, s, \cdot), \hat{G}(\cdot, Y, s)$ had a continuous extension to $\bar{U} \setminus \{(Y, s)\}$ with $\hat{G}(Y, s, \cdot) \equiv \hat{G}(\cdot, Y, s) \equiv 0$ on ∂U. We note that (a) of Lemma 2.2 holds with G replaced by $\hat{G}$ and c by a constant depending only on n as well as the constants in (1.2), (1.3) for $\hat{A}, \hat{B}$. In fact the proof of (a) of Lemma 2.2 in the smooth case for $r \approx y_0$ was given in (3.23), (3.24) of chapter I. The proof for other values of r is the same.

We shall essentially get G as a certain weak limit of smooth Green functions. However to prove $(b), (c)$ of Lemma 2.2 we need to carefully choose the sequence. For this purpose, let $A_j(X, t) = A(x_0 + j^{-1}, x, t), B_j(X, t) = B(x_0 + j^{-1}, x, t)$ for $j = 3, 4, \dots$, and $(X, t) \in \{(Y, s) : y_0 \geq -j^{-1}\}$. Then A_j, B_j satisfy (1.2) - (1.4) and for fixed j, B_j is essentially bounded by Mj. Now we can choose sequences of smooth functions which satisfy (1.2) - (1.4) (with uniform constants) and converge pointwise on U to A_j, B_j. Using the basic estimates in Lemmas 3.3 - 3.5 of chapter I we see for fixed $(Y, s) \in U$ and j that a subsequence involving smooth $\hat{G}(\cdot, \cdot)$ converges uniformly in a certain Hölder norm on $Q_{d_1}(X, t) \times Q_{d_2}(Y, s)$ to $G_j(\cdot, \cdot)$. Here $d_1, d_2, (X, t), (Y, s)$ are as in (e) of Lemma 2.2. We can also choose this subsequence so that for each $(Y, s) \in U$, the sequences involving $\hat{G}(\cdot, Y, s), \hat{G}(Y, s, \cdot)$ converge weakly in $L^2(-d^2 + \tau, d^2 + \tau, H^1_{\text{loc}}(\hat{Q}_d(Z, \tau))$ to $G_j(\cdot, Y, s), G_j(Y, s, \cdot)$, whenever $\bar{Q}_d(Z, \tau) \subset U \setminus \{(Y, s)\}$, where $\hat{E} = \{X : (X, t) \in E \text{ for some } t \in R\}$. From the above remarks we observe that (a) of Lemma 2.2 (with a constant depending only on γ_1, M, n) and $(d), (e)$ of Lemma 2.2 are valid for G_j (with an exponent depending only on γ_1, M, n). Now since $\|B_j\|_{L^\infty(U)} \le Mj$ we can further choose $\hat{A}, \hat{B}$ in our sequences so that (3.13) of chapter I holds uniformly in rectangles of small side length which touch ∂U. From this fact and the remark after (3.22) in chapter I we see that Lemma 3.9 of chapter I holds for $\hat{A}, \hat{B}$ with uniform Hölder exponent and constant. Using this fact, Harnack's inequality, and (a) of Lemma 2.2 for the Green's functions in the sequence, we deduce that (c) of Lemma 2.2 is valid for G_j with a constant which may depend on j.

We now use Lemmas 3.3 - 3.5 of chapter I to argue as in the above smooth case and get G satisfying $(a), (d), (e)$ of Lemma 2.2, as a certain weak limit of a subsequence of $\{G_j\}$. Let $\omega_j^*(X, t, \cdot) = \omega(x_0 + j^{-1}, x, t, \cdot)$, for $(X, t) \in U. j = 3, 4, \ldots,$. Then ω_j^* is a weak solution to (1.1) in U relative to A_j, B_j and from (2.1) as well as (a) of Lemma 2.2 we find

$$(2.3) \qquad G_j(\cdot, Y, s) \le c\, y_0^{-n}\, \omega_j^*(\cdot, Q_{r_1}(z, \tau))$$

on $\partial Q_{y_o/100n}(Y, s)$ whenever $y_0 \ge j^{-1}, (z, \tau) \in Q_{y_0/16}(y, s)$, and $y_0/4 \le r_1 < y_0$. From (c) of Lemma 2.2 for G_j we see that $G_j(\cdot, Y, s)$ has continuous boundary value zero. Using this fact, (2.3), and the maximum principle (see the remark after Lemma 3.38 of chapter I) for solutions to (1.1) we conclude that (2.3) holds in $U \setminus Q_{y_0/100n}(Y, s)$ with a constant independent of $j = 1, 2, \ldots,$. Letting $j \to \infty$ in (2.3) we get (b) of Lemma 2.2. (c) of Lemma 2.2 is a consequence of (b), Lemma 3.2 for A, B, of chapter II and the fact that $\omega(\cdot, Q_{r_1}(z, \tau))$ has continuous boundary value zero at points of $R^n \setminus \bar{Q}_{r_1}(z, \tau)$. The proof of Lemma 2.2 is now complete. $\square$

Next suppose that either $\Omega = U$ or $\Omega = Q_d(X, t) \cap U$ for some $d > 0, (X, t) \in U$ and also that $v \in L^2(-T, T, H^1_{\text{loc}}(\hat{U})), 0 < T < \infty$, satisfies

$$(2.4) \qquad \int_U [\langle A^* \nabla v + F, \nabla \xi \rangle - (B^* \nabla v + f)\xi - v\xi_s]\, dY ds = 0$$

whenever $\xi \in C_0^\infty(\Omega)$ and A^*, B^* satisfy (1.2)-(1.4). Here $F = (F_0, \ldots, F_{n-1})$ and $f, F_i \in L^2(U)$ for $0 \le i \le n - 1$. Also f, F_i each vanish outside of a compact set $K \subset U$. We say that v is a weak solution to $L^* v = \nabla \cdot F + f$ in Ω where L^* is as in (1.1) relative to A^*, B^*. Let $d_1 < d_2$ and $\bar{Q}_{d_2}(Y, s) \subset \Omega$. We shall need the

Cacciopoli inequality:

$$\int_{Q_{d_1}(Y,s)} |\nabla v|^2 \, dZ d\tau$$

(2.5)
$$\leq c \int_{Q_{d_2}(Y,s)} [\, |F|^2 + (d_2 - d_1)^2 f^2 \,] \, dZ \, d\tau$$

$$+ \, c \, (d_1 - d_2)^{-2} \int_{Q_{d_2}(Y,s)\setminus Q_{d_1}(Y,s)} v^2 \, dZ d\tau.$$

Here c depends on n, the distance of $\bar{Q}_{d_2}(Y,s)$ to $\partial\Omega$ and the constants in (1.2)-(1.3). If $d_2 \leq y_0/2$, then c can be chosen to depend only on n and the constants in (1.2), (1.3). (2.5) follows easily from (2.4) by using Cauchy's inequality with ϵ's and routine juggling, once it is shown that $ve^{-\lambda t}$ times a certain cutoff function can essentially be used as a test function (see [A]). Let O be an open set with $K \subset O$ and $\bar{O} \subset U$. By a solution to the Cauchy problem $Lu = \nabla \cdot F + f$, $u|_{\partial U} = g$, where g is continuous and bounded on ∂U, we mean a function $u \in L^2(-T, T, H^1_{loc}(\hat{U}))$ for $0 < T < \infty$ which is bounded outside of O, satisfies (2.4) in U (with A^*, B^* replaced by A, B) and is continuous on $\bar{U}$ with $u = g$ on ∂U. With this terminology we prove

Lemma 2.6. *Let A, B, ω, G be as in Lemma 2.2 and f, F as in (2.4). Then for a given continuous, bounded g on ∂U, the Cauchy problem, $Lu = \nabla \cdot F + f$, $u|_{\partial U} = g$ has a unique solution:*

$$u(X,t) = \int_{\partial U} g \, d\omega(X,t,\cdot) + \int_U [-\langle F, \nabla_Z G(X,t,\cdot) \rangle + f \, G(X,t,\cdot)] \, dZ d\tau.$$

Proof: Put $u^*(X,t) = \int_{\partial U} g \, d\omega(X,t,\cdot)$ and

$$u^{**}(X,t) = \int_U [-\langle F, \nabla_Z G(X,t,\cdot) \rangle + f \, G(X,t,\cdot)] \, dZ d\tau$$

for $(X,t) \in U$. We first consider u^{**}. Replace G, f, F by $\hat{G}, \hat{f}, \hat{F}$ in the definition of u^{**}, where $\hat{G}$ is as above, and call the resulting function $\hat{u}$. We assume that $\hat{f}, \hat{F}$ have support in $\bar{O}$. Then from Schauder type estimates we deduce for smooth $\hat{f}, \hat{F}$ that $\hat{u}$ is the unique solution to the Cauchy problem $\hat{L}\hat{u} = \nabla \cdot \hat{F} + \hat{f}$, $\hat{u}|_{\partial U} = 0$, where $\hat{L}$ is defined as in (1.1) relative to $\hat{A}, \hat{B}$. Choose $0 < d_1 < d_2, (Y,s)$ so that $\bar{O} \subset Q_{d_1}(Y,s)$ and $\bar{Q}_{d_2}(Y,s) \subset U$. Recall that Lemma 3.9 of chapter I held uniformly for the sequence we used to define G_j (with constants that could depend on j). Thus we can use $(a) - (c)$ of Lemma 2.2 and Lemma 3.3 of chapter I for $\hat{G}$ as well as (1.4) to conclude that $\hat{u}$ is uniformly bounded on $U \setminus Q_{d_1}(Y,s)$. Using this fact, (2.5), and essentially Poincare's inequality, we deduce that

(2.7) $\|\hat{u}\|_{L^2(Q_{d_1}(Y,s))} \leq c\,(\|f\|_{L^2(U)} + \|F\|_{L^2(U)})$.

Here c depends on $j, n, d_2 - d_1$, the distance of $Q_{d_2}(Y,s)$ to ∂U and the constants in (1.2)-(1.3). Next from (2.5), (2.7) we see that $\hat{u}$ is locally bounded in $L^2(-T, T, H^1_{loc}(\hat{U})), 0 < T < \infty$, with constants having the same dependence as c above. Approximating f, F by smooth $\hat{f}, \hat{F}$ whose supports are all contained in $\bar{O}$ and taking a weak limit as in Lemma 2.2 we get a solution $\tilde{u}_j$ to the Cauchy problem, $L_j \tilde{u}_j = \nabla \cdot F + f$, $\tilde{u}_j|_{\partial\Omega} = 0$, where L_j is defined as in (1.1) relative to A_j, B_j. From our construction of $G_j(\cdot, \cdot)$ as a certain weak limit and estimates using $(a) - (c)$ of Lemma 2.2, Lemma 3.3 of chapter I, it can be deduced that

$$(2.8) \quad \tilde{u}_j(X,t) = \int_U \left[-\langle F, \nabla_Z G_j(X,t,\cdot) \rangle + f\, G_j(X,t,\cdot) \right] dZ d\tau, \ (X,t) \in U.$$

Also from (2.3) and Lemma 3.3 of chapter I for G_j we see that we can use (2.5) for j large to get (2.7) with $\hat{u}$ replaced by $\tilde{u}_j$ and constants independent of j. Thus we can take a subsequence of the subsequence of j's used to define G so that the corresponding subsequence of $\{\tilde{u}_j\}$ converges weakly in $L^2(-T, T, H^1_{loc}(\hat{U}))$, $0 < T < \infty$, to $\tilde{u}$, a solution to the Cauchy problem $L\tilde{u} = \nabla \cdot F + f$, $\tilde{u}|_{\partial U} = 0$. Finally from (2.8), weak convergence, Lemma 3.3 of chapter I, and (a) of Lemma 2.2 it can be shown that $\tilde{u} = u^{**}$ almost everywhere on U. To complete our analysis of u^{**} we note from (c) of Lemma 2.2 and Lemma 3.3 for G that $u^{**}(X,t) \to 0$ as $x_0 \to 0$. As for u^*, we see by an easy functional analysis type argument, that this function is the solution to the Dirichlet problem for (1.1), A, B with boundary function g. Thus $u = u^* + u^{**}$ is the desired solution. Uniqueness of u follows from the maximum principle in Lemma 3.38 of chapter I. $\square$

Next we put

$$(2.9) \qquad E_j = E_j(X,t) = Q_{2^j x_0}(X,t) \setminus Q_{2^{j-1} x_0}(X,t) \text{ for } j = 0, \pm 1, \ldots,$$

and with this notation prove

Lemma 2.10. *Let f, F, v be as in (2.5) relative to $A, B, \Omega = Q_{4x_0}(X,t) \cap U$. Then*

$$\int_{E_0} |\nabla v|^2\, G(X,t,\cdot)\, dY ds \le c\, \|v\chi\|^2_{L^\infty(U)} + c\, \|v\chi\|_{L^\infty(U)} \int_U G(X,t,\cdot)\, |f|\, \chi\, dY ds$$

$$+ c\|v\chi\|_{L^\infty(U)} \left(\int_U |\nabla G(X,t,\cdot)|\, |F|\chi\, dY ds + c\, x_0^{-1} \int_U G(X,t,\cdot)\, |F|\chi\, dY ds \right)$$

$$+ c \int_U G(X,t,\cdot)\, |F|^2\, \chi\, dY ds,$$

where χ denotes the characteristic function of $Q_{2x_0}(X,t) \setminus Q_{x_0/4}(X,t)$.

Proof: Recall formula (3.6a) of chapter I for smooth $\hat{A}, \hat{B}, \hat{G}$. Taking a limit as above we see for $\phi \in C_0^\infty(R^{n+1})$ and fixed j, that

$$(2.11)$$
$$\phi(X,t) = \int_U \left[\langle A_j \nabla \phi, \nabla_Y G_j(X,t,\cdot) \rangle + G_j(X,t,\cdot)\, (\phi_s - B_j \nabla \phi) \right] dY ds$$

$$+ \int_{\partial U} \phi(y,s)\, d\omega_j(X,t,y,s),$$

whenever $(X,t) \in U$, where ω_j is parabolic measure defined relative to A_j, B_j. Let $Q = Q_{2x_0}(X,t) \setminus Q_{x_0/4}(X,t)$ and let $\theta \in C_0^\infty(Q)$ with

$$x_0 |\nabla\theta\|(Y,s) + x_0^2 |\tfrac{\partial}{\partial t}\theta|(Y,s) \le c(n) \text{ for } (Y,s) \in Q$$

while $\theta \equiv 1$ on $Q_{x_0}(X,t) \setminus Q_{x_0/2}(X,t)$. Next let κ be an even function in $C_0^\infty[(-1,1)]$ with $\int_\mathbb{R} \kappa\, dx = 1$ and first derivatives bounded by $c(n)$. Let $\kappa_\delta(\tau) = \delta^{-1} \kappa(\tau/\delta)$ for given $\delta > 0$ and if g is an integrable function on R let $g * \kappa_\delta$ denote convolution of g with κ_δ. If h is a real valued function defined on a subset of R^{n+1} we put $h_\delta(Y,s) = h(Y,\cdot) * \kappa_\delta(s)$ whenever the convolution makes sense. Finally we extend $G_j(X,t,\cdot)$ to a continuous function on $R^{n+1} \setminus \{(X,t)\}$ by putting $G_j(X,t,\cdot) \equiv 0$ in $R^{n+1} \setminus U$.

For v as in Lemma 2.10 we assume, as we may, that $\|v\chi\|_{L^\infty(U)} < \infty$ and put $v_j(Y,s) = v(y_0 + j^{-1}, y, s)$, $f_j(Y,s) = f(y_0 + j^{-1}, y, s)$, $F^j(Y,s) = F(y_0 + j^{-1}, y, s)$. Then for j large enough we find that

(i) v_j satisfies (2.4) relative to A_j, B_j, f_j, F^j in
$\Omega_j = \{(Z,\tau) : (Z,\tau) \in Q_{3x_0}(X,t) \text{ and } z_0 > -j^{-1}\}$,

(2.12) (ii) v_j is Hölder continuous in a neighborhood of $Q \cap \partial U$,

(iii) $|\nabla G_j|(X,t,\cdot)$ has square integrable distributional derivatives on $Q_{4x_0}(X,t) \setminus Q_{x_0/8}(X,t)$.

(ii) is a consequence of Lemma 3.4 of chapter I for v and the fact that f, F have compact support. (iii) follows from $\|B_j\|_{L^\infty(U)} \leq Mj$ and the same argument as in (3.18) of chapter I. Approximating by smooth functions and using (2.12) we see that the function ξ defined by

$$\xi = \begin{cases} [(v_j)_\delta \, \theta^2 \, G_j(X,t,\cdot)]_\delta & \text{on } U \cap Q, \\ 0 & \text{in } Q \setminus U \end{cases}$$

can be used as a test function in (2.4) for v_j. Putting this function in (2.4) we get after some rearranging that if $w = (v_j)_\delta$, and $\nabla = \nabla_Y$, then

(2.13)
$$I_1 = \int_Q \langle (A_j \nabla v_j)_\delta, \nabla[w\,\theta^2\,G_j(X,t,\cdot)]\rangle\, dY ds$$
$$= \int_Q (B_j \nabla v_j)_\delta \, w\,\theta^2\,G_j(X,t,\cdot)\, dY ds + \int_Q (f_j)_\delta \, w\,\theta^2\,G_j(X,t,\cdot)\, dY ds$$
$$- \int_Q \langle F^j_\delta, \nabla(w\,\theta^2\,G_j(X,t,\cdot))\rangle\, dY ds - \frac{1}{2}\int_Q \frac{\partial}{\partial s}(w^2)\,\theta^2\,G_j(X,t,\cdot)\, dY ds$$
$$= I_2 + I_3 + I_4 + I_5.$$

We note that
(2.14)
$$I_1 = \int_Q \langle A_j\nabla w, \nabla w\rangle\,\theta^2 G(X,t,\cdot)\, dY ds + \int_Q \langle A_j\nabla w, \nabla\theta^2\rangle\, wG(X,t,\cdot)\, dY ds$$
$$+ \frac{1}{2}\int_Q \langle A_j\nabla(w\,\theta)^2, \nabla G(X,t,\cdot)\rangle\theta^2\rangle\, dY ds - \int_Q \langle A_j\nabla\theta, \nabla G(X,t,\cdot)\rangle w^2\,\theta dY ds$$
$$+ \int_Q \langle (A_j\nabla v_j)_\delta - A_j\nabla w, \nabla[w\,\theta^2\,G_j(X,t,\cdot)]\rangle\, dY ds$$
$$= I_{11} + I_{12} + I_{13} + I_{14} + I_{15}.$$

Also

(2.15)
$$I_2 = \frac{1}{2} \int_Q B_j \nabla(w\,\theta)^2 \, G_j(X, t, \cdot) \, dY \, ds - \int_Q B_j \nabla\theta \, w^2 \theta \, G_j(X, t, \cdot) \, dY \, ds$$

$$+ \int_Q [(B_j \nabla v)_\delta - B_j \nabla w)] \, w\,\theta^2 \, G_j(X, t, \cdot) \, dY \, ds$$

$$= I_{21} + I_{22} + I_{23}.$$

From (1.2) we see that

(2.16)
$$\int_Q |\nabla w|^2 \, \theta^2 \, G(X, t, \cdot) \, dY \, ds \leq c \, I_{11}.$$

Using Cauchy's inequality with ϵ ' s we deduce that

(2.17)
$$|I_{12}| + |I_{14}| + |I_{22}|$$

$$\leq \frac{1}{4} I_{11} + c \, \|w^2 \chi\|_{L^\infty(U)}^2 \, x_0^{-1} \Big(\int_Q y_0^{-1} \, G_j(X, t, \cdot) + |\nabla \, G_j(X, t, \cdot)| \, dY \, ds \Big).$$

We also observe that

$$I_4 = - \int_Q \langle \, F_\delta^j \, , \, \nabla w \, \rangle \, \theta^2 \, G_j(X, t, \cdot) \, dY \, ds$$

$$- \int_Q \langle \, F_\delta^j, \nabla(\theta^2) \, \rangle \, w G_j(X, t, \cdot) \, dY \, ds - \int_Q w\theta^2 \langle \, F_\delta^j, \nabla G_j(X, t, \cdot) \rangle \, dY \, ds$$

$$= I_{41} + I_{42} + I_{43}.$$

Moreover,

(2.18)
$$|I_{41}| + +|I_{42}| + |I_{43}| \leq \frac{1}{4} I_{11} + c \int_Q |F_\delta^j|^2 \, \theta^2 \, G_j(X, t, \cdot) \, dY \, ds$$

$$+ c \, \|w\theta\|_{L^\infty(U)} \, [\int_Q |F^j| \, (x_0^{-1} \, G_j(X, t, \cdot) + \theta \, |\nabla G_j(X, t, \cdot)|) \, dY \, ds].$$

Next approximating by smooth functions and using (2.12) we deduce that (2.11) is valid with $\phi = (w\theta)^2$. From (2.11) and (a) of Lemma 2.2 we find that

(2.19)
$$| - I_{13} + I_{21} + I_5 | \leq c \, \|w\theta\|_{L^\infty(U)}^2.$$

Finally we note that the constants in (2.16)-(2.19) depend at most on the constants for A, B in (1.2)-(1.4). We first let $\delta \to 0$ and use (2.12) (iii) to conclude that $I_{15} + I_{23} \to 0$. Then we let $j \to \infty$ through values used in the sequence defining G.

We deduce from (2.13)-(2.19) that

(2.20)
$$\int_Q |\nabla v|^2 \, \theta^2 \, G(X,t,\cdot) \, dY ds$$

$$\leq c \, \|v^2 \, \chi\|^2_{L^\infty(U)} \, [\int_Q x_0^{-1} \, (\, y_0^{-1} \, G(X,t,\cdot) \, + \, |\nabla G(X,t,\cdot)| \,) \, dY ds\,]$$

$$+ \, c \, \|v\chi\|_{L^\infty(U)} \int_Q G(X,t,\cdot)|f| \, dY ds \, + \, c \int_Q |F| \, |\nabla G(X,t,\cdot)| \, dY ds$$

$$+ \, c \, x_0^{-1} \, \|v\chi\|_{L^\infty(U)} \int_Q |F| \, G(X,t,\cdot) \, dY ds \, + \, c \int_Q |F|^2 \, G(X,t,\cdot) \, dY ds\,.$$

Now from $(a) - (d)$ of Lemma 2.2 and Lemma 3.3 we see by the same argument as in (3.30) of chapter I that

$$\int_{Q_{y_0/16}(Y,s)} \theta \, (z_0^{-1} \, G(X,t,\cdot) \, + \, |\nabla G(X,t,\cdot)| \,) \, dZ d\tau \leq c \, z_0 \, \omega(X,t,Q_{z_0/2}(z,\tau)),$$

whenever $(z,\tau) \in Q_{y_0/16}(Y,s)$. This inequality implies that

$$\int_Q (\, z_0^{-1} \, G(X,t,\cdot) \, + \, |\nabla G(X,t,\cdot)| \,) \, dZ d\tau$$

(2.21)
$$\leq c \int_{Q_{6x_0}(X,t)} y_0^{-(n+1)} \, \omega(X,t,Q_{y_0/2}(y,s)) \, dY ds \, \leq c x_0.$$

In (2.21) the lefthand inequality is obtained from writing the first integral as a sum over Whitney rectangles and using the above inequality. The righthand inequality follows from interchanging the order of integration in the second integral. Putting (2.21) into (2.20) we conclude the validity of Lemma 2.10. □

Armed with Lemma 2.10 we are ready to begin the proof of the special case of Theorem 1.7 mentioned at the beginning of this section. To begin we assume only that A_1, B_1 satisfy (1.5) $(a),(b)$ and not necessarily (c) of this display. That is,

(a) The continuous Dirichlet problem for (1.1), A_1, B_1 always has a unique solution,

(b) If ω_1 denotes the corresponding parabolic measure, then $\lambda_1 \, \omega_1(d,x,t \mid 2d^2, Q_d(x,t)) \geq 1.$

Let G_1 be the Green's function corresponding to (1.1), A_1, B_1. We assume that

(2.22)
Either the backward Harnack inequality in Lemma 3.11 holds for G_1 or $A_2 \equiv A_1$.

Put

$$d \, \hat{\nu}(Y,s) = y_0^{-1} \text{ ess sup } \{[|A_2 - A_1|^2 \, + z_0^2 \, |B_2 - B_1|^2](Z,\tau) :$$
$$(Z,\tau) \in Q_{y_0/16}(Y,s)\} \, dY ds,$$

$$\xi(y,s) = \int_0^\infty \frac{d\hat{\nu}}{dY ds} \, (z_0,y,s) \, dz_0\,.$$

We prove

Lemma 2.23. *Let A_i, B_i satisfy (2.1)-(2.3) for $i = 1, 2$. Suppose that A_1, B_1 satisfy (1.5)(a), (b), and that (2.22) is valid. There exists $\epsilon_3 > 0$ depending on $\gamma_1, M, n, \lambda_1$ such that if $\|\xi\|_{L^\infty(\mathbb{R}^n)} \leq \epsilon_3^2$, then the Dirichlet problem for (1.1) corresponding to g (continuous and bounded on ∂U), A_2, B_2 has a unique solution u given for $(X, t) \in U$ by*

$$u(X,t) = u_1(X,t) + \int_U \langle (A_2 - A_1)\nabla u, \nabla G_1(X,t,\cdot)\rangle \, dY\, ds$$

$$+ \int_U (B_2 - B_1)\, \nabla u\, G_1(X,t,\cdot) dY\, ds.$$

u_1 is the solution to the continuous Dirichlet problem for (1.1), A_1, B_1 with $u_1 \equiv g$ on ∂U. Moreover,

$$\|u - u_1\|_{L^\infty(U)} \leq c(\gamma_1, M, n, \lambda_1)\, \epsilon_3\, \|g\|_{L^\infty(\partial U)}.$$

Proof: Note that Lemma 2.23 does not make a Carleson measure asuumption on B_1, B_2 individually, as in Theorem 1.7. We first prove Lemma 2.23 under the assumption that

$(\dagger)$ There exist $\delta > 0$ such that $A_2 \equiv A_1$ and $B_2 \equiv B_1$ in $U \cap \{(Z,\tau) : z_0 \leq \delta\}$.

To show the existence of u satisfying the above integral equation, we use Picard iteration. Put

$$u_{k+1}(X,t) = u_1(X,t) + \int_U \langle (A_2 - A_1)\nabla u_k, \nabla G_1(X,t,\cdot)\rangle \, dY\, ds$$

$$+ \int_U (B_2 - B_1)\, \nabla u_k\, G_1(X,t,\cdot)\, dY\, ds$$

$$= u_1(X,t) + \int_U H_k(X,t,\cdot)\, dY\, ds,$$

for $k = 1, 2, \ldots$, whenever these integrals make sense. We write,

$$(u_2 - u_1)(X,t) = \sum_{j=-\infty}^{\infty} \int_{E_j} H_1(X,t,Y,s) dY\, ds$$

where E_j is as in (2.9) relative to (X, t). We note from $\|\hat\nu\| \leq \epsilon_3^2$ that

(2.24) $\|A_2 - A_1\|_{L^\infty(U)} + \|x_0\, (B_2 - B_1)\|_{L^\infty(U)} \leq c(n)\, \epsilon_3.$

We also note that the interior estimates in Lemmas 3.3 - 3.5 of chapter I hold for weak solutions to (1.1) in U corresponding to A_1, B_1, as we see once again from approximating such solutions by smooth solutions and taking limits. Let β be the exponent of Hölder continuity in Lemma 3.4 of chapter I corresponding to (1.1), A_1, B_1. If $j \leq -1$ we deduce from (2.24), $(a) - (c)$ of Lemma 2.2, and Lemmas 3.3,

3.4 of chapter I that

$$\int_{E_j} |H_1|(X,t,\cdot)\,dY\,ds \le c\,\epsilon_3 \int_{E_j} |\nabla u_1|\,[\,y_0^{-1}\,G_1(X,t,\cdot) + |\nabla G_1(X,t,\cdot)|\,]\,dY\,ds$$

$$\le c\epsilon_3 \left(\int_{E_j} |\nabla u_1|^2\,dY\,ds\right)^{1/2} \left(\int_{E_j} y_0^{-2}\,G_1^2(X,t,\cdot) + |\nabla G_1(X,t,\cdot)|^2\,dY\,ds\right)^{1/2}$$

$$\le c\,\epsilon_3\,2^{\beta j}\,\|u_1\|_{L^\infty(U)} \le c\,\epsilon_3\,2^{\beta j}\,\|g\|_{L^\infty(\partial U)},$$

where the last inequality is a consequence of the maximum principle in Lemma 3.38 of chapter I. Summing over $j \le -1$, it follows that

$$(2.25) \qquad \int_{Q_{x_0/2}(X,t)} |H_1|(X,t,\cdot)\,dY\,ds \le c\epsilon_3\,\|g\|_{L^\infty(\partial U)}\,.$$

If $j \ge 0$, put $X_j = (2^{j+5}\,x_0,\,x)$, $t_j = t + 4^{j+5}\,x_0^2$ for $j = 0,1,\dots,$. To avoid confusion we indicate the dependence of E_j on (X,t). We claim that

$$(2.26)$$

$$\int_{E_j(X,t)} |H_1|(X,t,\cdot)\,dY\,ds \le c\epsilon_3 2^{-j\alpha} \left(\int_{E_0(X_j,t_j)} |\nabla u_1|^2\,G_1(X_j,t_j,\cdot)\,dY\,ds\right)^{1/2}$$

where α is the exponent in Lemma 3.2 of chapter II relative to A_1, B_1. To prove (2.26) we put

$$a = a_r(Z,\tau) =, \min_{Q_r(Z,\tau)} \{|\frac{d\hat\nu}{dY\,ds}|^{1/2}(Y,s)\}$$

and note that if $(Z,\tau) \in E_j(X,t)$, $j = 0,1,\dots,$ then from (2.22), Lemma 2.2, Lemma 3.2 of chapter II, and Lemmas 3.5, 3.3 of chapter I, we get for $z_0/1000 \le r \le z_0/64$, that

$$(2.27)$$
$$\int_{Q_r(Z,\tau)} |(A_2 - A_1)\nabla u_1|\,|\nabla G_1(X,t,\cdot)|\,dY\,ds$$

$$\le caz_0^{(n+1)/2}\,G_1(X,t,Z,\tau - z_0^2/32^2) \left(\int_{Q_r(Z,\tau)} |\nabla u_1|^2\,dY\,ds\right)^{1/2}$$

$$\le ca2^{-j\alpha}\,z_0^{(n+1)/2}\,G_1(X_j,t_j,Z,\tau - z_0^2/32^2) \left(\int_{Q_r(Z,\tau)} |\nabla u_1|^2\,dY\,ds\right)^{1/2}$$

$$\le c2^{-j\alpha} \left(\int_{Q_r(Z,\tau)} y_0^{-1}G_1(X_j,t_j,\cdot)d\hat\nu\right)^{1/2} \left(\int_{Q_r(Z,\tau)} |\nabla u_1|^2 G_1(X_j,t_j,\cdot)dY\,ds\right)^{1/2}$$

Next we divide E_j into Whitney rectangles $\{\bar Q_{r_l}(Z_l,\tau_l)\}$ with disjoint interiors and such that if $Q_{\hat r}(\hat Z,\hat\tau) \in \{Q_{r_l}(Z_l,\tau_l)\}$, then $\hat z_0/1000 \le r \le \hat z_0/64$. Using (2.27) with $Q_r(Z,\tau)$ replaced by the above Whitney rectangles, summing and using Cauchy's inequality we obtain

$$\int_{E_j(X,t)} |(A_2 - A_1)\nabla u_1|\,|\nabla G_1(X,t,\cdot)|\,dY\,ds \;\le$$

(2.28)
$$c\,2^{-j\alpha}\left(\int_{E_0(X_j,t_j)} y_0^{-1}\,G_1(X_j,t_j,\cdot)\,d\hat{\nu}(Y,s)\right)^{1/2}$$

$$\cdot\left(\int_{E_0(X_j,t_j)} |\nabla u_1|^2\,G_1(X_j,t_j,\cdot)dY\,ds\right)^{1/2}.$$

We note that Lemma 2.2 (b) holds for $0 < r_1 \le y_0/4$, provided we allow c in this equality to also depend on r_1, as follows easily from the proof of Lemma 2.2. Using this new version of Lemma 2.2 (b) for $r_1 = r_1(n) > 0$ sufficiently small and interchanging the order of integration in the first integral on the righthandside of (2.28) we get

(2.29)
$$\int_{E_0(X_j,t_j)} y_0^{-1}\,G_1(X_j,t_j,\cdot)\,d\hat{\nu}(Y,s) \le c\int_{\mathbb{R}^n} \xi(y,s)\,d\omega_1(x,t_j,y,s)$$

$$\le c\,\epsilon_3^2.$$

Using (2.29) in (2.28) we see that

$$\int_{E_j(X,t)} |(A_2 - A_1)\nabla u_1|\,|\nabla G_1(X,t,\cdot)|dY\,ds$$

$$\le c\,2^{-j\alpha}\epsilon_3\left(\int_{E_0(X_j,t_j)} |\nabla u_1|^2\,G_1(X_j,t_j,\cdot)dY\,ds\right)^{1/2}.$$

By a somewhat easier argument we also find that

$$\int_{E_j(X,t)} |(B_2 - B_1)\nabla u_1|\,G_1(X,t,\cdot)dY\,ds$$

$$\le c\,2^{-j\alpha}\epsilon_3\left(\int_{E_0(X_j,t_j)} |\nabla u_1|^2\,G_1(X_j,t_j,\cdot)dY\,ds\right)^{1/2}.$$

From the above inequalities we deduce that claim (2.26) is true. From the fact that u_1 is a weak solution to $L_1 u_1 = 0$, Lemma 2.10, and the maximum principle we get with (X,t) replaced by (X_j,t_j),

(2.30)
$$\int_{E_0(X_j,t_j)} |\nabla u_1|^2\,G_1(X_j,t_j,\cdot)\,dY\,ds \le c\,\|g\|_{L^\infty(\partial U)}^2.$$

Using (2.30) in (2.26) and summing we obtain

(2.31)
$$\int_{U\setminus Q_{x_0/2}(X,t)} |H_1|\,G_1(X,t,\cdot)\,dY\,ds \le c\epsilon_3\,\|g\|_{L^\infty(\partial U)}.$$

Combining (2.31) and (2.25) we see that

(2.32)
$$\|u_2 - u_1\|_{L^\infty(U)} \le c\epsilon_3\,\|g\|_{L^\infty(\partial U)}.$$

Note from assumption $(+)$ and Lemma 2.6 that u_2 is the solution to the Cauchy problem $L_1 u_2 = \nabla \cdot F_1 + f_1$ where L_1 is the operator in (1.1) defined relative to A_1, B_1 and

$$f_1 = (B_2 - B_1)\nabla u_1$$

$$F_1 = (A_2 - A_1)\nabla u_1.$$

Using this note, Lemma 2.10, and (2.32) we see that u in $L^2(-T, T, H^1_{loc}(\hat{U}))$, $0 < T < \infty$. We also assert for $0 < r \le x_0/2$ and $(X, t) \in U$ that for some $c \ge 1$, we have

(2.33)

$$\Phi(r, u_2 - u_1) = r^{-n}\int_{Q_r(X,t)} |\nabla(u_2 - u_1)|^2 dY\,ds \le c^+ (r/x_0)^\beta \epsilon_3^2 \|g\|^2_{L^\infty(\partial U)},$$

where again β is the constant in Lemma 3.4 for weak solutions to (1.1), A_1, B_1. To prove (2.33) we observe from (2.5), (2.24), (2.32), and the above note that if $x_0/100 \le r \le x_0/2$, then

$$\Phi(r, u_2 - u_1) \le c\epsilon_3^2\, r^{-n}\int_{Q_{3r/2}(X,t)} |\nabla u_1|^2 \, dY\,ds + c\epsilon_3^2 \|g\|^2_{L^\infty(\partial U)}$$

$$\le c\epsilon_3^2 \|g\|^2_{L^\infty(\partial U)}.$$

Thus (2.33) holds for the above values of r. To prove this inequality for $0 < r \le x_0/100$, we write $u_2 - u_1 = C_1 + D_1$, where for $(Z, \tau) \in U$, $Q = Q_r(X, t)$, and $G_1 = G_1(Z, \tau, \cdot)$, we have

$$C_1(Z, \tau) = \int_Q \langle (A_2 - A_1)\nabla u_1, \nabla G_1 \rangle \, dY\,ds + \int_Q (B_2 - B_1)\nabla u_1\, G_1 \, dY\,ds,$$

$$D_1(Z, \tau) = \int_{U\setminus Q} \langle (A_2 - A_1)\nabla u_1, \nabla G_1 \rangle \, dY\,ds + \int_{U\setminus Q} (B_2 - B_1)\nabla u_1\, G_1 \, dY\,ds.$$

Arguing as in the display above (2.25) we see for some $c = c(\gamma_1, M, n) \ge 1$ that

$$|C_1(Z, \tau)| \le c\,\epsilon_3\,(r/x_0)^\beta \|g\|_{L^\infty(\partial U)}, \quad \text{for } (Z, \tau) \in U \setminus Q_{2r}(X, t).$$

Using this inequality, the fact that $\Phi(r, u_1) \le c(r/x_0)^{2\beta}$, and (2.5) for C_1 it follows that there exists $c^* = c^*(\gamma_1, M, n) \ge 1$ for which

(2.34) $$\Phi(r, C_1) \le c\,\epsilon_3^2\,(r/x_0)^{2\beta} \|g\|^2_{L^\infty(\partial U)}. \le c^*\,\epsilon_3^2\,(r/x_0)^\beta \|g\|^2_{L^\infty(\partial U)}.$$

Next we note from Lemma 2.6 that D_1 is a weak solution to (1.1) corresponding to A_1, B_1 in Q. We claim for $0 < 100r_1 \le r$, that there exists $\hat{c} = \hat{c}(\gamma_1, M, n) \ge 1$ such that

(2.35) $$\Phi(r_1, D_1) \le \hat{c}\,(r_1/r)^{2\beta}\,\Phi(r/2, D_1).$$

To prove this statement we observe from Lemmas 3.3, 3.4 of chapter I that if $r/8 < \rho < r/4$ and a denotes the average of D_1 on $Q_\rho(X, t)$, then

$$\Phi(r_1, D_1) \le c\,(r_1/r)^{2\beta}\, r^{-(n+2)}\int_{Q_\rho(X,t)} (D_1 - a)^2 \, dY\,ds.$$

So to complete the proof of (2.35) it suffices to show that the integral involving $D_1 - a$ times r^{-n-2} is less than or equal to $c\,\Phi(r/2, D_1)$. This estimate does not follow directly from Poincáre's inequality since the gradient of D_1 is only in the space variable. However choosing an appropriate test function in (2.5) with u replaced by D_1 it is not difficult to show for some ρ as above and $t - \rho^2 < s < t + \rho^2$ that if $a(s)$ denotes the average of D_1 with respect to Lebesgue n measure on $Q_\rho(X, t) \cap (R^n \times \{s\})$, then

$$|a(s) - a| \leq c\,\Phi(r/2, D_1).$$

This inequality and Poincáre's inequality applied in $Q_\rho(X, t) \cap (R^n \times \{s\})$ for $t - \rho^2 < s < t + \rho^2$ give the desired inequality. Thus (2.35) is valid. We now use (2.34) and (2.35) to prove (2.33). We shall show that if (2.33) holds for some $r, 0 < r \leq x_0/100$, with c^+ replaced by $\tilde{c}$, then for $\tilde{c}$ large enough there exists $\theta = \theta(\gamma_1, M, \beta, n), 0 < \theta < 1$, for which this inequality also holds at θr. That is,

(2.36)
$$\Phi(\theta r, u_2 - u_1) \leq \tilde{c}\,\epsilon_3^2\,(\theta r/x_0)^\beta\,\|g\|_{L^\infty(\partial U)}^2.$$

Iterating (2.36) (starting with $r = x_0/100$) and using the fact that $\Phi(\rho, u_2 - u_1) \leq c\Phi(r, u_2 - u_1)$, for $\theta r \leq \rho \leq r$, we get (2.33). Hence we need only prove (2.36). As for (2.36) we see from (2.34), (2.35) with $r_1 = \theta r$, that

$$\Phi(\theta r, u_2 - u_1) \leq 2[\Phi(\theta r, C_1) + \Phi(\theta r, D_1)]$$

$$\leq 2\theta^{-n}\,\Phi(r, C_1) + 2\,\hat{c}\,\theta^{2\beta}\,\Phi(r/2, D_1)$$

$$\leq (2\,\theta^{-n} + 2^{n+2}\,\theta^{2\beta}\,\hat{c})\,\Phi(r, C_1) + 2^{n+2}\,\hat{c}\,\theta^{2\beta}\,\Phi(r, u_2 - u_1)$$

$$\leq [(2\,\theta^{-n} + 2^{n+2}\,\hat{c})\,c^* + 2^{n+2}\,\hat{c}\,\theta^{2\beta}\,\tilde{c}]\,(r/x_0)^\beta\,\epsilon_3^2\,\|g\|_{L^\infty(\partial U)}^2.$$

We first choose θ so that $2^{n+2}\,\hat{c}\,\theta^\beta = 1/2$. We next put

$$\tilde{c} = 2\,\theta^{-\beta}\,(2\,\theta^{-n} + 2^{n+2}\hat{c})\,c^*.$$

From the above inequality we see for these values of $\tilde{c}, \theta$ that (2.36) is true. From the remark following (2.36) we conclude that assertion (2.33) is true.

Next we show that

(2.37)
$$\int_{E_0(X,t)} |\nabla(u_2 - u_1)|^2\,G_1(X, t, \cdot)dY ds \leq c\,\epsilon_3^2\,\|g\|_{L^\infty(\partial U)}^2.$$

To prove (2.37) we use (2.32) and Lemma 2.10 to write for $v_1 = u_2 - u_1$,

(2.38)
$$\int_{E_0(X,t)} |\nabla v_1|^2\,G_1(X, t, \cdot)dY ds \leq c\,\epsilon_3^2\,\|g\|_{L^\infty(\partial U)}^2$$

$$+ c\epsilon_3\,\|g\|_{L^\infty(\partial U)} \int_U G_1(X, t, \cdot)\,|f_1|\,\chi\,dY ds$$

$$+ c\epsilon_3\,\|g\|_{L^\infty(\partial U)} \left(\int_U [\,|\nabla G_1(X, t, \cdot)| + x_0^{-1}\,G_1(X, t, \cdot)]\,|F_1|\,\chi\,dY ds \right)$$

$$+ c \int_U G_1(X, t, \cdot)\,|F_1|^2\,\chi\,dY ds$$

$$= c\epsilon_3^2\,\|g\|_{L^\infty(\partial U)}^2 + c\epsilon_3\,\|g\|_{L^\infty(\partial U)}\,(T_1 + T_2) + c\,T_3,$$

where f_1, F_1, are as below (2.32). Arguing as in (2.27)-(2.29) and using (2.30) we get that

$$|T_1| + |T_2| \le c \int_{E_0(X,t)} |(B_2 - B_1)\nabla u_1| \, G_1(X,t,\cdot) \, dY \, ds$$

$$+ c \int_{E_0(X,t)} |(A_2 - A_1)\nabla u_1| \, [\, |\nabla G_1(X,t,\cdot)| \, + \, x_0^{-1} G_1(X,t,\cdot) \,] \, dY \, ds$$

$$\le c\,\epsilon_3 \left(\int_{E_0(X_1,t_1)} |\nabla u_1|^2 \, G_1(X_1,t_1,\cdot) \, dY \, ds \right)^{1/2} \le c\,\epsilon_3 \, \|g\|_{L^\infty(\partial U)} \, .$$

Also from (2.24) we have

$$|T_3| \le c\epsilon_3^2 \int_{E_0(X_1,t_1)} |\nabla u_1|^2 \, G_1(X_1,t_1,\cdot) \, dY \, ds \le c\,\epsilon_3^2 \, \|g\|_{L^\infty(\partial U)}^2 \, .$$

Putting these estimates for the T's into (2.38) we get (2.37).

We now proceed by induction. Put $v_k = u_{k+1} - u_k$ for $k = 1, 2, \ldots,$ and suppose for $1 \le k \le l$ that we have shown

$$(a) \qquad \|v_k\|_{L^\infty(U)} \le (c_1 \, \epsilon_3)^k \|g\|_{L^\infty(\partial U)},$$

$$(b) \qquad \Phi(r, v_k) \le (r/x_0)^\beta \, (c_1 \, \epsilon_3)^{2k} \|g\|_{L^\infty(\partial U)}^2$$
$$\text{for } (X,t) \in U \text{ and } 0 < r \le \, x_0/2,$$

(2.39)

$$(c) \qquad \int_{E_0(X,t)} |\nabla v_k|^2 \, G_1(X,t,\cdot) dY \, ds \le (c_1 \, \epsilon_3)^{2k} \, \|g\|_{L^\infty(\partial U)}^2$$
$$\text{whenever } (X,t) \in U.$$

Note from (2.32), (2.33), and (2.37) that (2.39) is valid when $k = 1$ for c_1 sufficiently large. Using the induction hypothesis, we shall prove that (2.39) holds when $k = l + 1$ provided $c_1 = c_1(\gamma_1, M, n, \lambda_1, p_1)$ is large enough (independent of l). We proceed as in the case $k = 1$. In fact from (2.39) (b) with $k = l$ we get as in the display above (2.25) that for some $\bar{c} \ge 1$

$$(2.40) \qquad \int_{Q_{x_0/2}(X,t)} |H_{l+1} - H_l|(X,t,\cdot) \, dY \, ds \le (\bar{c}\,\epsilon_3)\,(c_1\,\epsilon_3)^l \, \|g\|_{L^\infty(\partial U)}.$$

Also using (2.39) (c) for $k = l$ and arguing as in the proof of (2.27)- (2.31) we find that

$$(2.41) \qquad \int_{U \setminus Q_{x_0/2}(X,t)} |H_{l+1} - H_l|(X,t,\cdot) \, dY \, ds \le (\bar{c}\,\epsilon_3)\,(c_1\,\epsilon_3)^l \, \|g\|_{L^\infty(\partial U)}.$$

Combining (2.40) and (2.41) we get (2.39) (a) when $k = l+1$ provided $c_1 \ge \bar{c}$. We note from assumption $(+)$ and Lemma 2.6 that v_k is the solution to the Cauchy problem $L_1 v_k = \nabla \cdot F_k + f_k$ where L_1 is the operator in (1.1) defined relative to A_1, B_1 and for $k > 1$,

$$f_k = (B_2 - B_1)\nabla v_{k-1}$$

$$F_k = (A_2 - A_1)\,\nabla v_{k-1}.$$

For $k = 1$, f_1, F_1 are as defined earlier. Using this note, (2.5), and (2.39) for $k = l$, we see that v_{l+1} in $L^2(-T, T, H^1_{loc}(\hat{U}))$, $0 < T < \infty$. Moreover, using (2.5) and (2.39), we find as in (2.33) whenever $(X, t) \in U$ and $x_0/100 \le r \le x_0/4$,

$$\Phi(r, v_{l+1}) \le c\epsilon_3^2\, \Phi(3r/2, v_l) + (c^*\epsilon_3)^2\, (c_1\epsilon_3)^{2l}\, \|g\|^2_{L^\infty(\partial U)}$$

$$\le (c^{**}\epsilon_3)^2\, (c_1\epsilon_3)^{2l}\, \|g\|^2_{L^\infty(\partial U)}\,.$$

If $r \le x_0/100$, we put $v_{l+1} = C_{l+1} + D_{l+1}$, where for $(Z, \tau) \in U$, $Q = Q_r(X, t)$, and $G_1 = G_1(Z, \tau, \cdot)$, we have

$$C_{l+1}(Z, \tau) = \int_Q \langle (A_2 - A_1)\nabla v_l, \nabla G_1 \rangle\, dY ds + \int_Q (B_2 - B_1)\, \nabla v_l\, G_1\, dY ds,$$

$$D_{l+1}(Z, \tau) = \int_{U \backslash Q} \langle (A_2 - A_1)\nabla v_l, \nabla G_1 \rangle\, dY ds + \int_{U \backslash Q} (B_2 - B_1)\, \nabla v_l\, G_1\, dY ds.$$

Arguing as in the proof of (2.34) we deduce first from (2.39) with $k = l$ that

$$|C_{l+1}(Z, \tau)| \le c\, \epsilon_3\, (r/x_0)^{\beta/2}\, (c_1\,\epsilon_3)^l\, \|g\|_{L^\infty(\partial U)}, \text{ for } (Z, \tau) \in U \backslash Q_{2r}(X, t),$$

and second from this inequality, as well as (2.39), that

$$\Phi(r, C_{l+1}) \le c^*\, \epsilon_3^2\, (r/x_0)^\beta\, (c_1\,\epsilon_3)^{2l}\, \|g\|^2_{L^\infty(\partial U)}.$$

Using this inequality, (2.35) with D_1 replaced by D_{l+1}, and arguing as in the proof of (2.36) we obtain (2.39) (b) for $k = l+1$ with $(c_1\epsilon_3)^{2l+2}$ replaced by $c^*\, \epsilon_3^2\, (c_1\epsilon_3)^{2l}$. Since $(X, t) \in U$ is arbitrary and we can cover $Q_r(X, t)$, $x_0/4 \le r \le x_0/2$, by at most $c(n)$ rectangles of side length $x_0/8$, we conclude that (2.39) (b) holds for c_1 large enough. Finally to prove (2.39)(c) for $k = l + 1$, we use (2.40), (2.41), and Lemma 2.10 to get as in (2.38) that

$$\int_{E_0(X,t)} |\nabla v_{l+1}|^2\, G_1(X, t, \cdot) dY ds \le (c_-\epsilon_3)^2\, (c_1\epsilon_3)^{2l}\, \|g\|^2_{L^\infty(\partial U)}$$

$$+ (c_-\,\epsilon_3)^2\, (c_1\epsilon_3)^l\, \|g\|_{L^\infty(\partial U)} \left(\int_{E_0(X_1,t_1)} |\nabla v_l|^2\, G_1(X_1, t_1, \cdot) dY ds \right)^{1/2}$$

$$\le (c_-\,\epsilon_3)^2\, (c_1\epsilon_3)^{2l}\, \|g\|^2_{L^\infty(\partial U)},$$

where the last inequality was obtained from (2.39)(c) for $k = l$. Thus (2.39)(c) is true for $c_1 \ge c_-$ when $k = l + 1$ and so (2.39) is true. By induction we conclude that (2.39) holds for k a positive integer.

From (2.39) we see that $u = \lim_{k \to \infty} u_k$ exists in $L^2(-T, T, H^1_{loc}(\hat{U})) \cap L^\infty(U)$ for $0 < T < \infty$. From the definition of u_k and (+) it follows that u is the solution to the integral equation in Lemma 2.23. Note from this equation, (+), and Lemma 2.6 that u is a weak solution to $L_1 u = \nabla \cdot F + f$ where $f = (B_2 - B_1)\nabla u$ and $F = (A_2 - A_1)\nabla u$ or equivalently u is a weak solution to $L_2 u = 0$ where L_2 is defined as in (1.1) relative to A_2, B_2. Moreover from the definition of u_1 and (c) of Lemma 2.2 we conclude first that $u - u_1$ has limit zero at each point in ∂U and second that u_2 is a solution to the continuous Dirichlet problem for L_2 with boundary function g. Uniqueness of u is a consequence of the maximum principle in Lemma 3.38. Next we observe from (2.39) that for ϵ_3 sufficiently small, we have

$$\|u - u_1\|_{L^\infty(U)} \leq 2c_1 \epsilon_3 \|g\|_{L^\infty(\partial U)}.$$

Hence Lemma 2.23 is true under assumption $(+)$. To conclude the proof of Lemma 2.23 we approximate A_2, B_2 by $\hat{A}_j, \hat{B}_j, j = 1, 2, \ldots$, where $\hat{A}_j \to A_2$, $\hat{B}_j \to B_2$ pointwise as $j \to \infty$. Furthermore, $\hat{A}_j, \hat{B}_j$ satisfy the hypotheses of Lemma 2.23 and $(+)$ holds for $j = 1, 2, \ldots$. For example, if $(X, t) \in U$ let

$$\hat{A}_j(X, t) = \begin{cases} A_2(X, t) \text{ for } x_0 \geq j^{-1} \\ A_1(X, t) \text{ for } x_0 < j^{-1} \end{cases}$$

and define $\hat{B}_j$ similarly for $j = 1, 2, \ldots,$. From Lemma 2.23 we deduce first that parabolic measure can be defined for each member of the above sequence. Second we deduce that the conclusion of Lemma 2.23 is valid with A_2, B_2 replaced by $\hat{A}_j, \hat{B}_j$ and with constants independent of j. From this deduction, (1.5)(b) for ω_1, and the definition of parabolic measure we find that parabolic measure corresponding to each member of the above sequence satisfies (1.5) (b) with constants independent of j. Using this fact, Lemma 3.2 of chapter II and the same argument as in Lemma 3.37 of chapter I we get after letting $j \to \infty$ that the continuous Dirichlet problem for (1.1), A_2, B_2 has a solution. Again uniqueness follows from the maximum principle in Lemma 3.38. Finally it is easily checked that the last inequality in Lemma 2.23 is still valid. $\square$

Let ω_2 be parabolic measure defined relative to A_2, B_2, Next suppose in addition to (1.5) $(a), (b)$ that ω_1 satisfies (1.5)(c). That is

$$(c) \qquad \|\tfrac{d\omega_1}{dyds}(d, x, t + 2d^2, \cdot)\|_{\alpha_{p_1}(Q_d(x,t))} < \lambda_1 < \infty.$$

Using Lemma 2.23 we can now easily prove the following special case of Theorem 1.7.

Lemma 2.42. *Let A_i, B_i be as in Lemma 2.23 for $i = 1, 2$, and suppose also that (1.5)(c) holds for $\omega_1, \lambda_1, p_1 > 1$. Then (1.5) also holds for ω_2 and some $\lambda_2 = \lambda_2(\gamma_1, M, n, \lambda_1, p_1, \epsilon_3)$, $p_2 = p_2(\gamma_1, M, n, \lambda_1, p_1, \epsilon_3) > 1$. Also if $q \geq p_2/(p_2 - 1)$, then the $L^q(\mathbf{R}^n)$ Dirichlet problem always has a unique solution in the sense of (I) and (II) of Theorem 1.13.*

Proof: To prove Lemma 2.42 we observe as above that the conclusion of Lemma 2.23 holds with u_1, u replaced by $\omega_1(\cdot, F), \omega_2(\cdot, F)$, whenever F is a Borel measurable subset of ∂U. We note that the reverse Hölder assumption on ω_1 in (1.5) (c) implies (3.1) of chapter II for ω_1. That is, there exist $\eta_0, \eta_1 > 0$ depending only on $\gamma_1, M, \lambda_1, n, p_1$ such that if $E \subset Q_r(y, s)$ and E is Borel, then

$$|E|/|Q_r(y, s)| \geq 1 - \eta_0 \Rightarrow \omega_1(r, y, s + 2r^2, E) \geq \eta_1.$$

From (3.1) of chapter II for ω_1 and the conclusion of Lemma 2.23 for ω_1, ω_2, we deduce that ω_2 also satisfies (3.1) of chapter II with constants $\eta_0, \eta_1/2$, provided ϵ_3 is small enough. Using this fact and Lemma 3.6 of chapter II we find that ω_2 also satisfies (1.5) with constants having the same dependence as η_0, η_1. The last sentence in Lemma 2.23 (solution and uniqueness of the $L^q(\mathbf{R}^n)$ Dirichlet problem for $q \geq p_2/(p_2 - 1)$) follows from (1.5) for A_2, B_2, ω_2, and Lemma 4.19 of chapter

II. The proof of Lemma 2.42 is now complete. $\square$

Remark We write ω_i for $\omega_i(d, x, t + 2d^2, \cdot)$ when $i = 1, 2$ and $d > 0$, $(x, t) \in R^n$ are fixed. If ω_1, ω_2 are doubling measures, then from Lemma 2.23, [CF], and Lemma 3.10 of chapter I it is easily seen that ω_1, ω_2 are A^∞ weights with respect to each other on $Q_d(x, t)$. In the general case when neither ω_1 or ω_2 may be doubling, we can use Lemma 2.23 and mimic the proof of Lemma 3.6 of chapter II with Lebesgue measure replaced by ω_1. Doing this one can first show that ω_2 is absolutely continuous with respect to ω_1 on $Q_d(x, t)$ so $d\omega_2 = f \, d\omega_1$. Second if $Q_{2r'}(y', s') \subset Q_d(x, t)$, then one can prove that there exists, $\theta = \theta(\gamma_1, M, n, \lambda_1, \epsilon_3) > 0$, such that

$$\int_{Q_{r'}(y', s')} f^{1+\theta} \, d\omega_1 \leq c\,\omega_1(d, x, t + 2d^2, Q_{2r'}(y', s'))^{-\theta} \left[\int_{Q_{2r'}(y', s')} f \, d\omega_1 \right]^{1+\theta},$$

where c depends on the same quantities as θ and also on the ratio of $\omega_1[Q_{2r'}(y', s')]$ to $\omega_1[Q_{r'/2}(y', s' + 2(r')^2)]$. We omit the details.

3. Proof of Theorem 1.7

As in section 2 we shall need to do some preliminary investigations before we can begin the proof of Theorem 1.7, in general. To this end suppose as in section 2 that the continuous Dirichlet problem corresponding to A, B (satisfying (1.2)-(1.4)) always has a unique solution and (1.2) holds for ω, the corresponding parabolic measure. That is, for some constant $c^* \geq 1$ we have

$$(3.1) \qquad\qquad c^* \omega(d, x, t + 2d^2, Q_d(x, t)) \geq 1$$

whenever $(x, t) \in R^n$ and $d > 0$. Let $F \subset R^n$ be a compact set and recall from section 2 of chapter II (see (2.19)) the definition of the parabolic distance function $\hat{\sigma}(\cdot, F)$. Let $\psi = \theta \hat{\sigma}(\cdot, F)$ where $\theta > 0$ and let

$$\rho(X, t) = (x_0 + P_{\gamma x_0} \psi(x, t), x, t), \ (X, t) \in U,$$

be as in sections 2-4 of chapter II, where γ is chosen so small that this mapping is one to one from U onto $\rho(U) \subset U$. If u satisfies (1.1) relative to A, B, then from (2.19), (2.27), (2.28) of chapter II we see once again that $\tilde{u} = u \circ \rho$ is a weak solution to (1.1) for some $\tilde{A}, \tilde{B}$ satisfying (1.2)-(1.4). We shall prove that

Lemma 3.2. *Let A, B, ω, ρ be as above. Then the continuous Dirichlet problem corresponding to (1.1), $\tilde{A}, \tilde{B}$ always has a unique solution. Moreover if $\tilde{\omega}$ denotes the corresponding parabolic measure, then for some $\tilde{c} = \tilde{c}(\gamma_1, M, \theta, c^*, n) \geq 1$ we have*

$$\tilde{c}\,\tilde{\omega}(d, x, t + 2d^2, Q_d(x, t)) \geq 1,$$

whenever $(x, t) \in R^n$ and $d > 0$.

Proof: Let $\tilde{A}_j(X, t) == \tilde{A}(x_0 + j^{-1}, x, t)$, $\tilde{B}_j(X, t) = \tilde{B}(x_0 + j^{-1}, x, t)$, for $j = 1, 2, \ldots$, and all $(X, t) \in U$. We first show that estimates similar to the ones in Lemma 3.9 of chapter I are valid for $\tilde{A}_j, \tilde{B}_j, j = 1, 2, \ldots$, with constants that are independent of j. To this end let $\tilde{u}_j$ be a solution to (1.1) corresponding to $\tilde{A}_j, \tilde{B}_j$ in $(0, 2r) \times Q_{2r}(y, s)$ with $\tilde{u}_j$ vanishing continuously on $Q_{2r}(y, s)$. We observe that if $Q_{3r/2}(y, s) \subset U \setminus F$, then $|\tilde{B}| \leq cM/r$ on $(0, \frac{5}{4}r) \times Q_{\frac{5}{4}r}(y, s)$. From this observation and the remark after (3.22) of chapter I we conclude that in this case Lemma 3.9 of chapter I is valid with u replaced by $\tilde{u}_j$ and constants independent of j. Otherwise, let $h = \omega(\cdot, Q_{3r/2}(y,s) \setminus Q_{9r/8}(y,s))$, $\tilde{h} = h \circ \rho$, and put $\tilde{h}_j(X,t) = \tilde{h}(x_0 + j^{-1}, x, t)$ for $j = 1, 2, \ldots$, and $(X,t) \in U$. We observe that $\tilde{h}_j$ is a solution to (1.1) corresponding to $\tilde{A}_j, \tilde{B}_j$ in U. Moreover $\tilde{h}_j \geq 0$ on $Q_{2r}(y,s)$ and $c'\,\tilde{h}_j \geq 1$ on $J = \partial[(0, \frac{5}{4}r) \times Q_{\frac{5}{4}r}(y,s)]$ thanks to (3.1) and Harnack's inequality. Then from the maximum principle (see the remark after Lemma 3.38 of chapter I) and Lemma 3.2 of chapter II for A, B, h we have for $(\hat{X}, \hat{t}) \in (0, r) \times Q_r(y,s)$ that

$$
\begin{aligned}
\tilde{u}_j(\hat{X}, \hat{t}) \;&\leq\; c\,(\max_J \tilde{u}_j)\,\tilde{h}_j(\hat{X}, \hat{t}) \\[4pt]
&\leq\; c\,\big(\max[j^{-1}, \hat{x}_0, \hat{\sigma}(\hat{x}, \hat{t}, F)]/r\big)^{\alpha}\,.
\end{aligned}
$$

(3.3)

If

$$
\hat{x}_0 \;\leq\; (100n)^{-1}\hat{\sigma}(\hat{x}, \hat{t}, F) \;=\; \hat{r}
$$

we observe that $|\tilde{B}| \leq cM/\hat{r}$ on $(0, 4\,\hat{r}) \times Q_{4\hat{r}}(\hat{x}, \hat{t})$. Thus if $I = \partial[(0, 2\hat{r}) \times Q_{2\hat{r}}(\hat{x}, \hat{t})]$ we can again use the remark after (3.22) of chapter I to conclude that

$$
\tilde{u}_j(\hat{X}, \hat{t}) \leq c\,(\max_I \tilde{u}_j)\,(\max[j^{-1}, \hat{x}_0]/\hat{r})^{\alpha}\,.
$$

Combining this inequality with (3.3) of chapter I we get

$$
\tilde{u}_j(\hat{X}, \hat{t}) \leq c\,(\max_J \tilde{u}_j)\,(\max[j^{-1}, \hat{x}_0]/r\,)^{\alpha}\,.
$$

Using the above inequality we can now argue as in the proof of Lemma 3.37 of chapter I (see $(i) - (iv)$ of this lemma) to get first that the continuous Dirichlet problem corresponding to $\tilde{A}, \tilde{B}$ has a unique solution and second that Lemma 3.9 of chapter I is valid with A, B replaced by $\tilde{A}, \tilde{B}$. From Lemma 3.9 with $u = 1 - \tilde{\omega}$ we see that Lemma 3.2 is true. $\square$

Next we put

$$
H(X,t) = \operatorname{ess\,sup}\,\{[y_0^{-1}|A_1 - A_2|^2 + y_0|B_2 - B_1|^2](Y,s) : (Y,s) \in Q_{x_0/2}(X,t)\}
$$

$$
L(X,t) = \operatorname{ess\,sup}\,\{[y_0^{-1}|A_1 - A_2|^2 + y_0|B_2 - B_1|^2](Y,s) : (Y,s) \in Q_{x_0/16}(X,t)\}
$$

$$
d\nu^*(X,t) = H(X,t)\,dX\,dt
$$

when $(X,t) \in U$. As in section 6 we shall need the following lemma.

Lemma 3.4. *Let A_1, B_1, A_2, B_2 satisfy (2.1)-(2.3) and (2.22). Suppose also that the continuous Dirichlet problem corresponding to (1.1), A_1, B_1 always has a unique solution and that (1.5) holds for ω_1. If $0 < \epsilon_4 < \epsilon_3$ is small enough (depending on $\gamma_1, M, n, p, \lambda, \epsilon_3$) and $\nu^*[(0,d) \times Q_d(x,t)] \leq \epsilon_4 |Q_d(x,t)|$, then there exists $\eta_0 = \eta_0(\epsilon_4)$, $\eta_1 = \eta_1(\epsilon_4)$, $0 < \eta_0, \eta_1 < 1/2$, such that the following statement is true. Let $u, 0 \leq u \leq 2$, be a solution to (1.1) in U, corresponding to A_2, B_2 which is continuous on $\bar{U}$. If $u \equiv 1$ on some closed set $E \subset Q_d(x,t)$ with*

$$
|E| \geq (1 - \eta_0)\,|Q_d(x,t)|,
$$

then

$$
u(d, x, t + 2d^2) \geq \eta_1.
$$

Proof: We note that Lemma 3.1 is similar to Lemma 2.1 of chapter II. In fact we can repeat essentially verbatim the argument in case (*) of Lemma 2.1 in chapter II to get F closed, $F \subset Q_d(x,t)$ such that if

$$\delta = \epsilon_4^{1/[100(n+2)]},$$

$$\Omega = \{(Z,\tau) \in U : z_0 > \hat{\sigma}(z,\tau,F)\},$$

then for ϵ_4 sufficiently small we have

(3.5)
$$|Q_d(x,t) \setminus F| \leq \delta |Q_d(x,t)|$$

and

(3.6)
$$\int_{\hat{\sigma}(z,\tau,F)}^{d/2} L(z_0,z,\tau)\,dz_0 \leq \delta \text{ whenever } (z,\tau) \in Q_d(x,t).$$

Let $A' = A_2, B' = B_2$ on $\Omega \cap [(0,d/2) \times Q_d(x,t)]$ and $A' = A_1$, $B' = B_1$, otherwise. Then from (3.6) we see that the hypotheses of Lemma 2.42 are satisfied for ϵ_4 sufficiently small. Thus the continuous Dirichlet problem corresponding to (1.1), A', B' has a unique solution and if ω' denotes the corresponding parabolic measure, then (1.5) holds for ω' provided λ, p are replaced by $\lambda', p' > 1$. Let $\psi = \hat{\sigma}(\cdot,F)$ and define ρ relative to ψ as in the display following (3.1). Then ρ maps U one to one and onto Ω. Let u' be a weak solution to (1.1), A', B' and put $\hat{u} = u' \circ \rho$. Then $\hat{u}$ is a weak solution to (1.1) corresponding to some $\tilde{A}, \tilde{B}$ satisfying (1.2)-(1.4). From our remarks on A', B' we see that Lemma 3.2 can be used with A, B replaced by A', B' to conclude that the continuous Dirichlet problem corresponding to (1.1), $\tilde{A}, \tilde{B}$ has a unique solution. Let $\tilde{u}$ denote the solution to the continuous Dirichlet problem for $\tilde{A}, \tilde{B}$ with boundary values $\tilde{u} = u \circ \rho$ on ∂U. we claim that if $Q_r(y,s) \subset Q_d(x,t)$, and $|Q_r(y,s) \setminus (E \cap F)| \leq 4\delta |Q_r(y,s)|$, then for some $c_- \geq 1$ we have

(3.7)
$$\tilde{u}(r,y,s+2r^2) \geq \tilde{\omega}(r,y,s+2r^2,E \cap F) \geq c_-^{-1}$$

provided $\epsilon_4 > 0$ is small enough. Here c_- has the same dependence as ϵ_4 in Lemma 3.2. To prove (3.7) we use Lemma 3.2 and argue as in the proof of (2.22). Choose $E' \subset E \cap F$ with $E' \subset Q_r(y,s)$ closed and with $|Q_r(y,s) \setminus E'| \leq 2|Q_r(y,s) \setminus (E \cap F)|$. Let $\{\bar{Q}_j\} \subset Q$ be a Whitney decomposition of $Q_r(y,s) \setminus E'$ into parabolic rectangles with side length $s(Q_i)$ in the space variables satisfying $(300)^{-n} \hat{\sigma}(Q_i,F) \leq s(Q_i) \leq 100^{-n}\hat{\sigma}(Q_i,F)$ for each i. Then from (1.5) for ω' and the maximum principle we see that if $\hat{Q}_j$ denotes the rectangle with the same center as Q_j and twice the side length in the space and time directions, and if

$$\omega^*(Z,\tau,K) = \omega'(\rho(Z,\tau),K), \ K = \text{ Borel set } \subset R^n,$$

then $c\,\omega^*(\cdot,\hat{Q}_j) \geq 1$ on Q_j. Also this function is a weak solution to (1.1) relative to $\tilde{A}, \tilde{B}$ so from the maximum principle we have

$$c\,\omega^*(\cdot,\hat{Q}_j) \geq \tilde{\omega}(\cdot,Q_j).$$

Using the reverse Hölder condition for ω' (i.e (1.5)(c)) in a now well known way and Lemma 3.2 for $\tilde{\omega}, Q_r(y,s)$, we conclude for some $\bar{c} \geq 1$ and $\delta > 0$ small enough that if $\hat{r} = r(1 + c\delta^{1/(n+1)})$, then

$$\tilde{\omega}(r, y, s + 2r^2, Q_r(y, s) \setminus (E \cap F)) \le c \sum_j \omega^*(r, y, s + 2r^2, \hat{Q}_j)$$

$$\le c\, \omega^*(r, y, s + 2r^2, Q_{\hat{r}}(y, s) \setminus F') \le \tfrac{1}{2}\tilde{\omega}(r, y, s + 2r^2, Q_r(y, s)).$$

From this inequality and Lemma 3.2 we get the righthand inequality in claim (3.7). The lefthand inequality follows from the definition of parabolic measure and the boundary values of u.

Next we note that $\tilde{u}, u \circ \rho$ satisfy the same pde in $(0, d/16) \times Q_{d/2}(x, t)$ and have the same continuous boundary values on ∂U. Using this fact, Lemma 3.2, and Lemma 3.2 of chapter II we find the existence of $c = c(\epsilon_4) \ge 1$, such that if $r = d/c$, then

(3.8) $$c_- |\tilde{u} - u \circ \rho|(r, y, s + 2r^2) \le 1/2,$$

where c_- is as in (3.7). Combining (3.8), (3.7) and using Harnack's inequality we conclude the validity of Lemma 3.4. $\square$

Extrapolation Revisited. To get Theorem 1.7 we shall essentially repeat the argument in section 4 of chapter II. We prove

Lemma 3.9. *Remove (2.22) from the statement of Lemma 3.4 and replace ϵ_4, ν^* by K, ν in this lemma, where ν is as defined above (1.6). Then this amended version of Lemma 3.4 remains valid whenever $0 < K < \infty$, provided $\eta_i = \eta_i(K, \gamma_1, M, \lambda_1, p_1, n)$, are defined suitably for $i = 0, 1$.*

Proof: Note that Lemma 3.9 is similar to Lemma 4.1 in chapter II. We claim that it suffices to prove Lemma 3.9 under the assumption that one of $(*), (**)$ are valid

$(*)$ $\qquad A_1 \equiv A_2, \; B_2 \equiv 0,$

$(**)$ $\qquad B_1 \equiv 0.$

Indeed to prove Lemma 3.9 in general we can first use $(*)$ to reduce the proof of this lemma to the situation when $B_1 \equiv 0$. We then obtain Lemma 3.9 from $(**)$. We continue under the assumption that either $(*)$ or $(**)$ holds and observe in either case that (2.22) is valid (thanks to Lemmas 3.11 and 3.14 of chapter I). We put

$$\tilde{H}(X, t) = \operatorname*{ess\,sup}\{[y_0^{-1}|A_1 - A_2|^2 + y_0(|B_2| + |B_1|)^2](Y, s) \\ : (Y, s) \in Q_{x_0/2}(X, t)\}$$

$$\tilde{L}(X, t) = \operatorname*{ess\,sup}\{[y_0^{-1}|A_1 - A_2|^2 + y_0(|B_2| + |B_1|)^2](Y, s)$$

$$: (Y, s) \in Q_{x_0/16}(X, t)\}$$

when $(X, t) \in U$ and note that $d\nu(X, t) = \tilde{H}(X, t)\, dX dt$. As in Lemma 4.1 of chapter II, we shall prove Lemma 3.9 by an induction type argument on K. From the remark following $(**)$ and Lemma 3.4, we see that Lemma 3.9 is valid for $K \le \epsilon_4(\gamma_1, M, \lambda_1, p_1, n)$. Suppose that whenever $\gamma_1, M, \lambda_1, p_1$ are given as above we have shown that Lemma 3.9 holds for $K \le K^*$ and $K^* \ge \epsilon_4(\gamma_1, M, \lambda_1, p_1, n)$

where $K^* = K^*(\gamma_1, M, \lambda_1, p_1, n)$. We assume as we may that M, λ_1, are both ≥ 100. We then put

(3.10)

$$\eta = \left[\frac{\epsilon_2(\gamma_1, M, \lambda_1, p_1, n)}{(\lambda_1 + M)(1 + K^*)c_1(n)} \right]^{2^{30n}}$$

$$\delta = \left[\frac{\epsilon_2(\gamma_1, M, \lambda_1, p_1, n)}{(\lambda_1 + M)(1 + K^*)c_1(n)} \right]^{20}$$

and shall show for $c_1 = c_1(n) \geq 1$ large enough that Lemma 3.9 is valid for $K \leq (1 + \eta)K^*$ provided $\eta_i = \eta_i(K)$, $i = 0, 1$ are defined suitably for $K^* < K \leq (1 + \eta)K^*$. We then get Lemma 3.9 by induction. To this end choose N such that $2^{-(N+1)} \leq \delta^5 \leq 2^{-N}$ and suppose first that

(3.11)
$$\int_{2^{-2N}d}^{d} \left(\int_{Q_d(x,t)} \tilde{H}(Z,\tau)\, dz d\tau \right) dz_0 \geq \eta K^* |Q_d(x,t)|.$$

Then as in the argument after (4.3) of chapter II we get that Lemma 3.9 holds for $K^* < K \leq (1 + \eta)K^*$ provided $\eta_0(K) \leq c_2(n)^{-1} \delta^{10(n+1)} \eta_0(K^*)$, $\eta_1(K) \leq c_3(\gamma_1, \delta, M, n)^{-1} \eta_1(K^*)$ and c_2, c_3 are large enough.

Next suppose that (3.11) is false. We again divide $Q_d(x,t)$ into subrectangles by the bisection method. Let G_m be the closed rectangles obtained in the m th subdivision for $m = 1, 2, \ldots,$. Then the rectangles in G_m have disjoint interiors and side length $2^{1-m}d$, $2^{1-2m}d^2$ in the space and time variables respectively. Let S_m be the subcollection of rectangles $Q_{2^{-m}d}(y,s)$ in G_m with

(3.12)

$$\int_{2^{-(N+j)d}}^{d} \int_{Q_{2^{-j}d}(y,s) \cap Q_d(x,t)} \tilde{H}(Z,\tau)\, dz d\tau dz_0$$

$$\leq (100n)^{100n^2} \eta K^* |Q_{2^{-j}d}(y,s)|$$

$$= \hat{\eta} K^* |Q_{2^{-j}d}(y,s)|$$

for $j = 1, 2, \ldots, m - 1$, while

(3.13)
$$\int_{2^{-(N+m)d}}^{d} \int_{Q_{2^{-m}d}(y,s) \cap Q_d(x,t)} \tilde{H}(Z,\tau)\, dz d\tau dz_0 \geq \hat{\eta} K^* |Q_{2^{-m}d}(y,s)|$$

Using the fact that (3.11) is false and a Calderòn-Zygmund type argument, we get as in section 4 of chapter II, a family of closed rectangles, $S = \bigcup S_m$ with disjoint interiors. Moreover if $(y,s) \notin \bigcup_{Q \in S} Q$, then (3.12) holds for $j = 1, 2, \ldots,$. Put $F^* = Q_d(x,t) \setminus \left(\bigcup_{Q \in S} Q \right)$. We consider two cases : $(a) |F^*| \geq 2\eta |Q_d(x,t)|$ and $(b) |F^*| < 2\eta |Q_d(x,t)|$.

If (a) holds, we suppose $\eta_0(K) \leq \eta/2$ for $K^* < K \leq (1 + \eta)K^*$ and set $d_1 = d[1 - \frac{\eta}{4(n+1)}]$. We observe that there exists F closed, $F \subset F^* \cap E \cap Q_{d_1}(x,t)$ with $|F| \geq \eta |Q_d(x,t)|$. Next we use (3.12) and argue as in previous proofs (see (4.6)-(4.7) of chapter II) to get

(3.14)
$$\int_{\delta\hat{\sigma}(z,\tau,F)}^{d} \tilde{L}(z_0, z, \tau)\, dz_0 \leq \delta$$

as well as

(3.15) $z_0 \, \tilde{L}(Z,\tau) \leq \delta$ on $\Omega \cap [(0, 3d/4) \times Q_d(x,t)]$.

Set

$$\psi = \delta \hat{\sigma}(\cdot, F),$$

$$\Omega = \{(Z,\tau) : z_0 > \psi(z,\tau)\},$$

$$\rho(Z,\tau) = (z_0 + P_{\gamma z_0}\psi(z,\tau), z, \tau),$$

when $(Z,\tau) \in U$. If (*) holds in the display below Lemma 3.9, we put $A' = A_2, B' = 0$ on Ω and $A' = A_1, B' = B_1$, otherwise in U. If (**) holds, let $A' = A_2, B' = B_2$ on $(0, d/2) \times Q_d(x,t)$ and $A' = A_1$ otherwise in U. From (3.14) and the observation following (**) we see that Lemma 3.4 can be applied with A_2, B_2 replaced by A', B'. We get that the corresponding parabolic measure, ω', satisfies (1.5) for some $\lambda', p' > 1$ and so also (3.1) of chapter II. Let u' be a solution to the Dirichlet problem for (1.1), A', B'. Then from (2.27) - (2.29) of chapter II,(3.15) and Lemma A in chaper I, we see that $u' \circ \rho$ satisfies (1.1) for some $\tilde{A}, \tilde{B}$ satisfying (1.2)-(1.4), as well as (3.13) of chapter I. Thus we can apply Lemma 3.37 of chapter I to obtain that the continuous Dirichlet problem corresponding to (1.1), $\tilde{A}, \tilde{B}$ has a unique solution and if $\tilde{\omega}$ denotes the corresponding parabolic measure, then $\tilde{\omega}$ is a doubling measure in the sense of this lemma. From this discussion we see that the hypotheses of Lemma 3.22 in chapter II are satisfied with ω_1, ω_2 replaced by $\omega', \tilde{\omega}$. From this lemma we find for some $c' \geq 1$, that

$$c' \, \tilde{\omega}(d, x, t + 2d^2, F) \geq 1.$$

Using this inequality we can now use the maximum principle and argue as above (4.9) of chapter II to deduce the existence of $c_4 \geq 1$, having the same dependence as η_0 such that if $0 < r < \eta d/c_3$, and $d_2 = d(1 - \frac{\eta}{8(n+1)})$, then

$$c' \, \tilde{\omega}(r, y, s, F) \geq 1 \text{ for some } (y,s) \in \bar{Q}_{d_2}(x,t).$$

Let $\tilde{u}$ be the weak solution to (1.1) corresponding to $\tilde{A}, \tilde{B}$ with $\tilde{u} = u \circ \rho$ on ∂U. Then since $u \geq 1$ on F it follows from the definition of $\tilde{\omega}$ that (3.7) holds with c_-^{-1} replaced by c' and (r, y, s) as above. Next we observe that $u \circ \rho, \tilde{u}$ satisfy the same pde in $(0, d/8) \times Q_d(x,t)$, as we see from the definition of A', B' and (2.27)-(2.29) of chapter II. Since $\tilde{u} - u \circ \rho$ vanishes on ∂U we can apply Lemma 3.2 of chapter II to conclude that (3.8) holds for all $(y,s) \in \bar{Q}_{d_2}(x,t)$ with c_- replaced by c' and $r \leq \eta d/c$, where c has the same dependence as η_0. Combining (3.8), (3.7) it follows that Lemma 3.9 is valid in case (a) when $K^* \leq K \leq (1 + \eta)K^*$.

Next we consider case (b). Arguing as in the proof of claim (4.11) of chapter II we first get a finite subcollection S' of S such that if $Q \in S'$ and $\eta' = \frac{\hat{\eta}}{(100n)^{10n}}$, then

(a) $\int_0^{2^{-(N+1)} s(Q)} \int_Q H(Z,s) \, dz d\tau dz_0 \leq (1 - \eta') K^* |Q|,$

(3.16)   (b) $\sum_{Q \in S'} |Q| \geq \eta' |Q_d(x,t)|,$

(c) $\hat{\sigma}(Q, Q') \geq 4n \max\{s(Q), s(Q')\}.$

If $\eta_0(K) \leq \eta_0(K^*)\,\eta^{n+2}$ for $K^* < K \leq (1+\eta)K^*$, then arguing as in the proof of (4.16) of chapter II we get the existence of a finite subset $\hat{S} = \{\, Q_{r_i}(y_i, s_i)\,\}_1^l$ of S' and $Q_{8r_i'}(z_i, \tau_i) \subset Q_{r_i}(y_i, s_i)$ such that for $1 \leq i \leq l$, we have $\frac{1}{2}\eta \leq \frac{r_i'}{r_i} \leq \eta$ and

$$(i) \qquad \int_{(0, r_i') \times Q_{r_i'}(z_i, \tau_i)} H(Z, \tau)\, dZ d\tau \;\leq\; K^* |Q_{r_i'}(z_i, \tau_i)|,$$

$$(ii) \qquad |E \cap Q_{r_i'}(z_i, \tau_i)| \;\geq\; (1 - \eta_0(K^*))|Q_{r_i'}(z_i, \tau_i)|,$$

$$(3.17) \qquad (iii) \qquad \sum_{i=1}^l |Q_{r_i'}(z_i, \tau_i)| \;\geq\; \eta^{n+2}\,|Q_d(x, t)|,$$

$$(iv) \qquad \text{Either } r \geq \eta d/100n \text{ for some } Q_r(y, s) \in \hat{S} \text{ or}$$
$$\cup_{Q \in \hat{S}}\, Q \subset Q_{d_1}(x, t).$$

First suppose there exists $Q_{r_i}(y_i, s_i) \in \hat{S}$ with $r_i \geq \eta d/100n$. Then from (3.17) $(i), (ii)$ and the induction hypothesis we see that $c\,u(r_i', z_i, \tau_i + 2(r_i')^2) \geq 1$ for some c having the same dependence as η_0. From this inequality and Harnack's inequality we conclude that Lemma 3.9 is valid for $K^* < K \leq (1+\eta)K^*$. Thus we assume that the second alternative in (3.17) (iv) occurs.

Put $F_+ = \{(y_i, s_i) : Q_{r_i}(y_i, s_i) \in \hat{S}\}$ and let

$$\tilde{\sigma}(z, \tau) = \begin{cases} r_i \text{ when } |z - y_i| + |\tau - s_i|^{1/2} \leq r_i,\ 1 \leq i \leq l, \\ = \hat{\sigma}(z, \tau, F_+), \text{ otherwise in } R^n. \end{cases}$$

Define ψ, Ω, ρ as following (3.15) with $\hat{\sigma}(\cdot, F)$ replaced by $\tilde{\sigma}$. Then from (3.12) we see that (3.14), (3.15) are valid. Next we define A', B' relative to $A_1, B_1, A_2, B_2, \Omega$ as in case (a). From (3.14), (3.15) we see that once again we can use Lemma 3.4 to get that ω' satisfies (1.5). We define $\tilde{A}, \tilde{B}$ relative to A', B', ρ as in in case (a). We then get that Lemma 3.37 of chapter I holds for $\tilde{\omega}$. Next from (3.16) (c) and the definition of $\tilde{\sigma}$ we see that Lemma 3.33 of chapter II can be applied with $K = \bigcup_{Q \in \hat{S}} \bar{Q}$. Applying this lemma we get for some $c'' \geq 1$ having the same dependence as η_0,

$$(3.18) \qquad c''\tilde{\omega}(d, x, t + 2d^2, \cup_{Q \in \hat{S}} \bar{Q}) \;\geq\; 1.$$

Now as earlier in case (b) we see from (3.17) and the induction hypothesis that for each i we have $c\,u(r_i', z_i, \tau_i + 2(r_i')^2) \geq 1$. Using this fact and Harnack's inequality we conclude that there exists $c^* \geq 1$, for which $c^*\,u \circ \rho \geq 1$ on $Q_{r_i'}(z_i, \tau_i + 4(r_i')^2)$, $1 \leq i \leq l$. Let $\tilde{u}$ be as in case (a). Then from the previous inequality for $u \circ \rho$ we deduce that

$$(3.19) \qquad c^*\,\tilde{u} \;\geq\; \tilde{\omega}(\cdot, \cup Q_{r_i'}(z_i, \tau_i + 4(r_i')^2))\,.$$

Also from (3.18) for $\tilde{\omega}$ and Lemma 3.37 (β) of chapter I we find for some $c^{**} \geq 1$ with the same dependence as η_0 that

$$(3.20) \qquad c^{**}\,\tilde{\omega}(d, x, t + 2d^2, \cup Q_{r_i'}(z_i, \tau_i + 4(r_i')^2))\;\geq\;1.$$

Using (3.19), (3.20) we can now argue as at the end of case (a) to get first (3.7) with F replaced by $\bigcup Q_{r_i'}(z_i, \tau_i + 4(r_i')^2)$ and then (3.8). We put $\eta_0(K) = \eta_0(K^*)\,\eta^{n+2}$ and for this value of η_0 conclude from (3.7), (3.8) as in case (a) that Lemma 3.9 is true when $K^* < K \le (1 + \eta)K^*$. By induction we obtain Lemma 3.9. $\square$

Proof of Theorem 1.7. We now prove Theorem 1.7. First we apply Lemma 3.9 to find that (3.1) of chapter II holds whenever $d > 0$ and $Q_d(x, t) \subset R^n$. From (3.1) and Lemma 3.6 of chapter II we deduce the validity of Theorem 1.7. $\square$.

Remark. We conjecture that Theorem 1.7 remains valid with ν replaced by ν^*. That is Theorem 1.7 is valid if instead of assuming $x_0(|B_1|^2 + |B_2|^2)\,dX\,dt$ is a Carleson measure we assume only that $x_0(|B_1 - B_2|^2)\,dX\,dt$ is a Carleson measure. In fact if we could prove that $\tilde{\omega}$ as in Lemma 3.2 is always a doubling measure, then it is easily seen that this stronger verion of Theorem 1.7 is valid. If one is unable to prove doubling for such parabolic measures, then another way to prove this conjecture would be to generalize Lemmas 3.22, 3.33 of chapter II to nondoubling measures and also to do away with assumption (2.22) in Lemma 2.23. Finally we note from section 1 that the elliptic verion of the above conjecture is true as we shall show in section 4.

4.   ELLIPTIC RESULTS

In this section we prove our elliptic results. For ease of notation we shall always assume that $n > 2$. We first prove some basic estimates similar to those in Lemma 2.2. To begin let A, B satisfy (1.2)-(1.4) with (X, t) replaced by X in $\hat{U} = \{X : x_0 > 0\}$ and suppose that the continuous Dirichlet problem for A, B, relative to the pde

$$(4.1) \qquad\qquad \nabla \cdot (A\nabla u) + B\nabla u = 0$$

always has a unique weak solution. If ω denotes the corresponding elliptic measure we assume for some positive $c^* < \infty$ that

$$(4.2) \qquad\qquad c^*\, \omega(d, x, B_d(x)) \ge 1$$

whenever $d > 0$ and $x \in R^n$. Here $B_d(x) = \{y : |y - x| < d\}$. We shall also denote the ball in R^n of radius d about X by $B_d(X)$ when there is chance of confusion. First we prove an analogue of Lemma 2.2.

Lemma 4.3. *Let A, B be as above and suppose that ω satisfies (4.2). There exists $G : \hat{U} \times \hat{U} \to R$ with the following properties. If $X, Y \in \hat{U}$, $X \ne Y$, and $r = |X - Y|$, then for some $c \ge 1$, $0 < \theta < \frac{1}{2}$ (depending only on γ_1, M, n, c^*), we have*

(a) $c^{-1} r^{2-n} \le G(X, Y) \le cr^{2-n}\ 0 < r < y_0/2,$

(b) $G(X, Y) \le cy_0^{2-n}\omega(X, B_{r_1}(z))$ *for* $z \in B_{y_0/16}(y), r > y_0/2, y_0/4 < r_1 < y_0,$

(c) $G(X, Y) < c\,(x_0/r)^\theta\, G(\hat{X}, Y)$ *for* $r > x_0$ *and* $\hat{X} = (x_0 + r, 0, \dots, 0),$

(d) $G(\cdot, Y)$ *is a weak solution to (4.1) and $G(Y, \cdot)$ is a weak solution to*
   $\nabla \cdot [\, A^\tau \nabla G(Y, \cdot) - B\, G(Y, \cdot)\,] = 0$ *in* $U \setminus \{Y\}$,

(e) *If $0 < d_1, d_2 < \min\{\, r/100n,\, y_0/2,\, x_0/2\,\}$, then $G(\cdot, \cdot)$ is Hölder continuous on $B_{d_1}(X) \times B_{d_2}(Y)$, with exponent independent of r_1, r_2, X, Y.*

Proof: The proof of this lemma is somewhat different than in the parabolic case so we include some details. Let $\hat{G}$ denote the Green's function defined relative to a 'smooth' $\hat{A}, \hat{B}$ satisfying (1.2)-(1.4) with (X, t) replaced by X. Then from Schauder's theorem, (1.4), and the divergence theorem applied in $\hat{U} \setminus \bar{B}_r(X)$, $r \leq x_0/2$, we deduce for some $c = c(\gamma_1, M, n) \geq 1$ that

$$1 = \int_{\mathbb{R}^{n-1}} |\nabla \hat{G}|^{-1} \langle \nabla \hat{G}, \hat{A} \nabla \hat{G} \rangle (X, 0, y) dy$$

$$= \int_{\partial B_s(X)} |X - Y|^{-1} \langle X - Y, \hat{A} \nabla_Y \hat{G}(X, Y) \rangle d\sigma(Y).$$

Here $d\sigma$ denotes surface area and $0 < s \leq x_0/2$. Integrating this inequality over $s \in (r/16, r/8)$, using the elliptic analogue of Lemma 3.3 in chapter I, as well as Hölder's and Harnack's inequalities, we get for some $c = c(\gamma_1, M, n) \geq 1$ that

(4.4) $$1 \leq c|X - Y|^{2-n} \hat{G}(X, Y) \text{ for } 0 < |X - Y| \leq x_0/2.$$

We note that (4.4) is just the reverse of the inequality obtained in the parabolic case by a similar argument.

As in section 2 we put $A_j(X) = A(x_0 + j^{-1}, x)$, $B_j(X) = B(x_0 + j^{-1}, x)$ for $j = 3, 4, \ldots$, and $X \in \{Y : y_0 \geq -j^{-1}\}$. Then A_j, B_j satisfy (1.2) - (1.4) and for fixed j, B_j is essentially bounded by Mj. Clearly we can choose sequences of smooth $\hat{A}, \hat{B}$ which converge pointwise to A_j, B_j on $\hat{U}$ and satisfy (1.2)-(1.4) with uniform constants. We can also choose this sequence so that (3.13) of chapter I holds (with (X, t) replaced by X) uniformly in cylinders of height and side length $\approx \epsilon_1 (Mj)^{-1}$ which touch ∂U. Thus an elliptic analogue of Lemma 3.10 in chapter I is valid (with uniform constants depending on j). Using elliptic analogues of the basic estimates in Lemmas 3.3 - 3.5 and Lemma 3.10 of chapter I, we see for fixed $Y \in \hat{U}$ and j that a subsequence involving smooth $\hat{G}(\cdot, \cdot)$ converges uniformly in a certain Hölder norm on $B_{d_1}(X) \times B_{d_2}(Y)$ to $G_j(\cdot, \cdot)$. Here d_1, d_2, X, Y are as in (e) of Lemma 4.3. Also we can choose this sequence so that for each $Y \in U$, the sequences involving $\hat{G}(\cdot, Y), \hat{G}(Y, \cdot)$ converge weakly in $H^1_{\text{loc}}(B_d(Z))$ to $G_j(\cdot, Y), G_j(Y, \cdot)$, whenever $\bar{B}_d(Z) \subset U \setminus \{Y\}$. From the above remarks and (4.4) we observe that the lefthand inequality in (a) of Lemma 4.3 is valid (with a constant depending only on γ_1, M, n) while from the elliptic version of Lemma 3.10 we see that the righthand inequality in (a) of Lemma 4.3 holds with a constant that in addition to the above quantities may also depend on j. Also $(d), (e)$ of Lemma 4.3 are valid for G_j (with an exponent depending only on γ_1, M, n). Next from the remark after (3.22) in chapter I we observe that the elliptic analogue of Lemma 3.9 holds in cylinders of height and radius $\approx \epsilon_1 (Mj)^{-1}$ with constants depending only on γ_1, M, n. That is, suppose $z \in R^{n-1}, 0 < d \leq \epsilon_1 (Mj)^{-1}$, and $u \geq 0$ is a weak solution to (4.1) in $(0, 2d) \times B_{2d}(z)$ (with A, B replaced by A_j, B_j), which vanishes continuously on $B_{2d}(z)$. Then there exists $\alpha = \alpha(\gamma_1, M, n) > 0$ and $c = c(\gamma_1, M, n)$ such that

(4.5) $$u(Y) \leq c (y_0/d)^\alpha u(d, z)$$

whenever $Y \in (0, d) \times B_d(z)$. From (4.5) and the same argument as in Lemma 3.37 of chapter I we see that the continuous Dirichlet problem correponding to A_j, B_j always has a unique solution. Let ω_j denote elliptic measure corresponding to A_j, B_j and set $\omega_j^*(Z, \cdot) = \omega(z_0 + j^{-1}, z, \cdot)$ when $Z \in \hat{U}$. Then ω^* is a weak solution to (4.1), with A, B replaced by A_j, B_j. From the maximum principle, (4.2),

Harnack's inequality and the definition of elliptic measure we have for some $c^{**} \geq 1$,

$$c^{**} \omega_j^*(\cdot, R^{n-1} \setminus B_{2d}(z)) \geq \omega_j(\cdot, R^{n-1} \setminus B_{2d}(z))$$

whenever $d \geq \epsilon_1 (Mj)^{-1}$. From this inequality and the elliptic analogue of Lemma 3.2 of chapter II for A, B we find for some $c = c(\gamma_1, M, n, c^*) \geq 1$ and $d \geq \epsilon_1 (Mj)^{-1}$ that

$$\omega_j(Y, R^{n-1} \setminus B_{2d}(z)) \leq c(y_0/d)^\alpha$$

for $Y \in (0, d) \times B_d(z)$. From (4.5) if follows that the above inequality actually holds whenever $0 < d < \infty$. Using this inequality for y_0 small enough, Harnack's inequality, and the fact that $\omega_j(\cdot, B_{2d}(z)) + \omega_j(\cdot, R^{n-1} \setminus B_{2d}(z)) = 1$, we conclude that (4.2) holds with ω replaced by ω_j. We can now use the elliptic version of Lemma 3.2 of chapter II to get that (4.5) holds for $j = 3, 4, \ldots$, with constants independent of j.

Next suppose that v is the solution to the continuous Dirichlet problem for (4.1), A_j, B_j with $v \equiv g$ on ∂U where $0 \leq g \leq 1$ is continuous on $\partial \hat{U}$ with support in $B_{2x_0}(x)$ and $g \equiv 1$ on $B_{x_0}(x)$. Then from (4.5), Lemma 3.2 of chapter II, and essentially Poincare's inequality we see for some $c' = c'(\gamma_1, M, n, c^*) \geq 1$, that

$$x_0^{n-2} \leq c' \int_{(x_0/c', x_0/2) \times B_{x_0}(x)} |\nabla v|^2 \, dZ.$$

From this inequality and the elliptic analogue of Lemma 2.10 for A_j, B_j we obtain that if m is the minimum of $G_j(X, \cdot)$ on $(x_0/c', x_0/2) \times B_{x_0}(x)$, then

$$m \, x_0^{n-2} \leq c' \int_{(x_0/c', x_0/2) \times B_{x_0}(x)} G_j(X, \cdot) |\nabla v|^2 \, dZ \leq c.$$

This inequality and Harnack's inequality imply that the righthand inequality in Lemma 4.3 (a) holds for $r = d/2$ with a constant independent of j. To obtain this inequality for other values of r one can use classical elliptic estimates similar to those in section 3. One proof for example would be to assume that the righthand inequality in (a) of Lemma 4.3 holds for $r = r_0 \leq y_0/2$ and some constant $\tilde{c}$. Let $G'(\cdot, Y)$ denote the Green's function for $B_{r_0}(Y)$ with pole at Y defined relative to (1.1), A_j, B_j. Then from properties of the Green's function similar to those listed in (3.6), (3.7) of chapter I we see that $G_j(\cdot, Y) \leq G'(\cdot, Y) + \tilde{c} r_0^{2-n}$ in $B_{r_0}(Y)$. Scaling to a ball of radius y_0 and using the same argument as above we find first that $G' \leq c r_0^{2-n}$ on $B_{r_0/2}(Y)$ and thereupon that the righthand inequality in (a) of Lemma 4.3 is true for $\tilde{c}$ large enough, whenever $r = r_0/2$. Thus (a) of Lemma 4.3 holds with constants independent of j. Using (a) of Lemma 4.3, letting $j \to \infty$, and arguing as in the proof of Lemma 2.2 we get G satisfying $(a) - (e)$ in Lemma 4.3. $\square$

Next we prove an elliptic analogue of Lemma 3.10 in chapter I.

Lemma 4.6 *Let A, B, ω, G be as in (4.1)-(4.2) and Lemma 4.3. If $B_{2r}(y) \subset B_d(x)$, then for some $c = c(\gamma_1, M, n, c^*) \geq 1$, we have*

$$c^{-1} r^{n-2} G(d, x, r, y) \quad \leq \omega(d, x, B_r(y))$$

$$\leq c \, \omega(d, x, B_{2r}(y)) \leq c^2 \, r^{n-2} G(d, x, r, y)$$

whenever $x \in R^{n-1}$ and $d > 0$.

Proof: Clearly the top inequality involving G, ω is implied by Lemma 4.3 (b). To prove the rest of this inequality we first show for given $\epsilon > 0$ the existence of $c(\epsilon) = c(\epsilon, \gamma_1, M, n, c^*) \geq 0$ such that

$$(4.7) \qquad \omega(\cdot, B_r(y)) \leq \epsilon\, \omega(\cdot, B_{2r}(y)) + c(\epsilon)\, r^{n-2}\, G(\cdot, r, y)$$

in $\hat{U} \setminus [\,(0, 3r/2) \times B_{r/2}(y)\,]$. The proof of (4.7) is essentially the same as the proof of (3.7) in chapter II except that instead of using $\omega_1(\cdot, E)$ we use $r^{n-2}\, G(\cdot, r, y)$ to make our comparisons, which is permissible thanks to the lefthand inequality in (a) of Lemma 4.3. We omit the details. $\square$

We note from (4.2) and (a) of Lemma 4.3 that Lemma 4.6 is true when $r \approx d$. Thus we assume $r/d = 2^{-N}$ for some large positive integer N. Iterating (4.7) we see that

$$\omega(d, x, B_r(y)) \leq \epsilon\, \omega(d, x, B_{2r}(y)) + c(\epsilon)\, r^{n-2}\, G(d, x, r, y)$$

$$\leq \epsilon^2\, \omega(d, x, B_{4r}(y)) + \epsilon\, c(\epsilon)\, (2r)^{n-2}\, G(d, x, 2r, y) + c(\epsilon)\, r^{n-2}\, G(d, x, r, y)$$

$$\leq \cdots \leq \epsilon^N \omega(d, x, B_{2d}(x)) + c(\epsilon) \Big[\sum_{i=1}^{N} \epsilon^{i-1} (2^{i-1} r)^{n-2} G(d, x, 2^{i-1} r, y) \Big] = T.$$

From Harnack's inequality and (a) of Lemma 4.3 we see for ϵ sufficiently small that there exists c_1 having the same dependence as the constant in Lemma 4.6 with $T \leq c_1\, r^{n-2}\, G(d, x, r, y)$. In view of the above inequality we conclude for some $c \geq 1$ that

$$\omega(d, x, B_r(y)) \leq c\, r^{n-2}\, G(d, x, r, y).$$

Thus $\omega(d, x, B_r(y)) \approx c\, r^{n-2}\, G(d, x, r, y)$ which along with Harnack's inequality implies Lemma 4.6. $\square$

Next let $\bar{\sigma}(\cdot, F)$ denote the Euclidean distance from the compact set $F \subset R^{n-1}$. Let A, B be as in (4.1) and put $\psi = \theta\, \bar{\sigma}(\cdot, F)$ where $\theta > 0$. Define ρ relative to ψ as in section 3 and note that if u is a weak solution to (4.1), corresponding to A, B, then $u \circ \rho$ is a weak solution to (4.1) corresponding to some $\tilde{A}, \tilde{B}$ satisfying (1.2)-(1.4) with (X, t) replaced by X. We prove

Lemma 4.8. *Let A, B, ω be as in (4.1), (4.2) and $\rho, \tilde{A}, \tilde{B}$ as above. Then the continuous Dirichlet problem corresponding to (1.1), $\tilde{A}, \tilde{B}$ always has a unique solution. Moreover if $\tilde{\omega}$ denotes the corresponding elliptic measure, then for some $\tilde{c} = \tilde{c}(\gamma_1, M, n, c^*) \geq 1$ we have*

$(a) \qquad \tilde{\omega}(r, y, B_r(y)) \geq 1, \text{ whenever } r > 0 \text{ and } y \in R^{n-1},$

$(b) \qquad$ *Lemma 4.6 holds with G, ω replaced by $\tilde{G}, \tilde{\omega}$, where $\tilde{G}$ is the Green's function corresponding to $\tilde{A}, \tilde{B}$.*

Proof: (a) of Lemma 4.8 is just the elliptic analogue of Lemma 3.2 and is proved in the same way as this lemma. (b) of Lemma 4.8 follows from (a), Lemma 4.3 for $\tilde{G}$, and Lemma 4.6. $\square$

Proof of Theorem 1.9 To prove Theorem 1.9 we repeat the argument in sections 2 and 3. Assumption (2.22) can now be done away with. Also assumptions $(*), (**)$ following (3.9) are unnecessary. In fact we needed $(*), (**)$ only to assure that (2.22) held and that certain parabolic measures obtained from using the ρ mapping were doubling so that Lemmas 3.22, 3.33 of chapter II could be applied. From Lemmas 4.6, 4.8 we see that the corresponding elliptic measures are always doubling so that we can use elliptic analogues of Lemmas 3.22, 3.33 in chapter II in our proof. In fact we can essentially just use the result in [DJK] mentioned earlier. Doing this we get Theorem 1.9. $\square$

Proof of Theorems 1.14, 1.15.   To prove Theorem 1.14 of chapter II we first consider elliptic pde's for which (1.2)-(1.4) hold in $\hat{U}$ and

(4.9) $A \in C^\infty \left(\bar{\hat{U}} \right)$ is lower triangular, $A_{00} \equiv 1, B \equiv 0$, and $y_0 |\nabla A(Y)| \leq \hat{\epsilon} \in \hat{U}$.

We shall show that if $x \in R^{n-1}, d > 0$, and $\hat{\epsilon} > 0$ is small enough, then $d\omega/dy\,(d, x, \cdot) \in \beta_2^*(B_d(x))$ where $\omega = \omega(d, x, \cdot)$ is elliptic measure corresponding to A. Indeed let $g(d, x, \cdot)$ be the corresponding adjoint Green's function with pole at (d, x). Then $\nabla \cdot (A^t \nabla_Y g(d, x, Y)) = 0$ when $Y \neq (d, x)$ where A^t is the transpose of A. We note that A^t is upper triangular and $A_{00}^t = A_{00} \equiv 1$. Differentiating the pde for $g = g(d, x, \cdot)$ with respect to y_l and using this note we see for $1 \leq l \leq n-1$ that

$$\nabla \cdot (A^t \nabla g_{y_l}) + \nabla \cdot (A_{y_l}^t \nabla g) = 0$$

where $A_{y_l}^t \nabla g$ does not involve any term in g_{y_0}. If $u = (g_{y_1}, \ldots, g_{y_{n-1}})$, then the above system has the same structure as (3.1) in chapter I and $u \equiv 0$ on $\partial \hat{U}$. It is easily seen for $\hat{\epsilon} > 0$ small enough that the argument in (3.15)-(3.17) of chapter I can be repeated with minor modifications to get the Cacciopoli inequality in (3.18) of chapter I. Using (3.18) for u, and Lemma 3.14 of chapter I we find for $z \in B_d(x)$ and $0 < r < d/4$ that

(4.10) $$\int_0^r \int_{B_r(x)} |\nabla u|^2 \, dY \leq cr^{-2} \int_0^{2r} \int_{B_{2r}(x)} |u|^2 \, dY \leq c\,r^{n-4} g^2(d, x, 3r, x).$$

Now we can use the pde for g to estimate $g_{y_0 y_0}$ pointwise in terms of $|u|, |\nabla u|$. Doing this, using (4.10), and making estimates by way of either Hardy's inequality or (3.16) of chapter I we see that

$$\int_0^r \int_{B_r(x)} |g_{y_0 y_0}|^2 \, dY \leq c \int_0^{2r} \int_{B_{2r}(x)} |\nabla u|^2 dY \leq c\,r^{n-4} g^2(d, x, 3r, x).$$

Finally from this inequality, Cauchy's inequality, (3.18) and Lemma 3.14 of chapter I for g, we conclude that

$$\int_{B_r(x)} |g_{y_0}|^2(0, y) \, dy \leq \int_{B_r(x)} |g_{y_0}|^2(r, y) \, dy$$

$$+ c\,r \int_0^r \int_{B_r(x)} |g_{y_0 y_0}|^2 \, dY + c\,r^{-1} \int_0^r \int_{B_r(x)} |g_{y_0}|^2 \, dY$$

$$\leq cr^{n-3} g^2(d, x, 3r, x) \leq cr^{1-n} [\omega(B_r(x))]^2.$$

This inequality and $d\omega/dy(0, y) \approx |\nabla g|(d, x, 0, y)$, for $y \in \partial \hat{U}$ are easily seen to imply that $d\omega/dy \in \beta_2^*(B_d(x))$. Thus Theorem 1.14 is valid in this case.

Next we use Theorem 1.9 and the above special case to deduce that Theorem 1.14 is valid when (1.2)-(1.4) hold, B satisfies (1.5) of chapter II, and A is as in (4.9). Moreover we observe that if u, B, A smooth satisfy $\nabla \cdot (A\nabla u) + B\nabla u = 0$

where A, B satisfy (1.2)-(1.7) of chapter II with $(X, t), dX dt$ replaced by X, dX then this equation can be rewritten in the form $\nabla \cdot (\tilde{A} \nabla u) + \tilde{B} \nabla u = 0$ where $\tilde{A}$ is lower triangular, $\tilde{A}_{00} \equiv 1$, and $\tilde{A}, \tilde{B}$ satisfy (1.2)-(1.7). Moreover if $x_0 |\nabla A|$ is small in $\hat{U}$, then so also is $x_0 |\nabla \tilde{A}|$. Thus Theorem 1.14 is vald when A, B are smooth,

(4.11) $\qquad A, B$ satisfy (1.2)-(1.7) and $x_0 |\nabla A|$ is sufficiently small in $\hat{U}$.

We remove the smoothness assumption on A, B by the same argument as the one after (4.4).

To continue the proof of Theorem 1.14 of chapter II, we prove an analogue of Lemma 2.1 in chapter II. In this lemma, $\hat{\mu}_1, \hat{\mu}_2$ are as in Theorem 1.14 of chapter II.

Lemma 4.12. *Let A, B satisfy (1.2)-(1.4) in $\hat{U}$ and suppose for some $x \in R^{n-1}, d > 0, \epsilon_1 > 0$ small that*

(i) $\quad$ *(1.7) of chapter II with $dX dt$ replaced by dX is valid in $(0, \infty) \times B_d(x)$.*

(ii) $\quad$ *$\tilde{\mu}[(0, d) \times B_d(x)] \leq \epsilon_1 |B_d(x)|$ where either $\tilde{\mu} = \hat{\mu}_1 + \hat{\mu}_2$ or (b) $\tilde{\mu} = \mu_1 + \mu_2$ and (**) of Theorem 1.10 in chapter II is valid.*

If $\epsilon_1 > 0$ is small enough (depending only on γ_1, M, n, Λ and possibly Λ_1), there exists $\eta_0 = \eta_0(\epsilon_1), \eta_1 = \eta_1(\epsilon_1), 0 < \eta_0, \eta_1 < 1/2$, such that the following statement is true. Let $u, 0 \leq u \leq 2$, be a solution to (1.1) in $\hat{U}$, corresponding to A, B as above, which is continuous on $\hat{U}$. If $u \equiv 1$ on some closed set $E \subset B_d(x, t)$ with

$$|E| \geq (1 - \eta_0) |Q_d(x, t)|,$$

then

$$u(d, x) \geq \eta_1.$$

Proof: Let $L(X) = x_0 |B(x)|^2 + x_0 |\nabla A|^2$ and if F ia closed, $F \subset B_d(x)$ set

$$\delta = \epsilon_1^{1/[1000(n+2)]}$$

$$\sigma(z, F) = \inf\{|z - y| : y \in F\}$$

$$\Omega = \{Z \in U : z_0 > \delta^4 \sigma(z, F).$$

We claim that for $\epsilon_1 > 0$ sufficiently small there exists F as above with

(+) $\qquad |B_d(x) \setminus F| \leq \delta |B_d(x)|$

(++) $\qquad z_0 L(Z) \leq \delta^{40}$, for $Z \in \Omega \cap [(0, 3d/4) \times B_d(x)]$

(4.13)

(+++) $\qquad \displaystyle\int_{\delta^4 \sigma(z, F)}^{3d/4} L(z_0, z) dz_0 \leq \delta^{100}$ for a.e $z \in B_d(x)$.

To prove (4.13) we can essentially copy the argument in Lemma 2.1 of chapter II except that now we do not have to worry about the integration by parts hypothesis (1.9) of chapter II or showing that A is near a constant matrix. We omit the details.

Let $P_\lambda(z) = \lambda^{-n} P(z/\lambda)$ where $P \in C_0^\infty(B_1(0))$ with $\int_{R^{n-1}} P(z) dz = 1$. Set $\rho(X) = (x_0 + P_\lambda \sigma(x, F), x)$. Using (4.13) and (2.27)- (2.31) of chapter II, we see that $u \circ \rho$ satisfies a pde with coeficients $\tilde{A}, \tilde{B}$ for which (1.2)-(1.7) of chapter II hold in $\hat{U}$ and $x_0|\nabla \tilde{A}|$ is small in $(0, d/2) \times B_{d/2}(x)$ when $\epsilon_1 > 0$ is small. Let $\phi, 0 \leq \phi \leq 1$, be in $C_0^\infty(-d/4, d/4)$ with $\phi = 1$ on $(-d/8, d/8)$. Let A_0 be the average of $u \circ \rho$ on $(d/8, d/4) \times B_{d/4}(x)$ and put $A_1(X) = (\tilde{A} - A_0)(X)\phi(x)\,\phi(x_0) + A_0$ while $B_1(X) = \tilde{B}(X)$ whenever $X \in \hat{U}$. Then A_1, B_1 satisfy (4.11) in $\hat{U}$ for $\epsilon_1 > 0$ small enough. Let u_1 be the solution to the Dirichlet problem for A_1, B_1 guaranteed by Theorem 1.9 with $u_1 = u \circ \rho$ on $\partial \hat{U}$. From (4.11) for A_1, B_1 we see that Lemma 4.12 is valid with u replaced by u_1. Lemma 4.12 for u_1 implies Lemma 4.12 for u as in (2.24) of chapter II. □

We can now extrapolate Theorem 1.14 from Lemma 4.12 just as Theorem 1.10 was extrapolated from Lemma 2.1 (see section 4 of chapter II). Again the argument is somewhat easier since we do not have to worry about the integration by parts hypothesis (1.9). □

Theorem 1.15 is deduced from Theorem 1.14 in the same way that Theorem 1.13 was deduced from Theorem 1.10. □

Remark. In Lemma 4.6 we showed that elliptic measure satisfying (4.2) is always doubling. We would not be surprised if a corresponding result held in the parabolic case. The chief obstacle to adapting the proof in [FS] to parabolic equations satisfying (1.1), (3.1) of chapter I is, as mentioned earlier, that we have not been able to obtain basic estimates for the adjoint pde in (3.1). Thus for example since constants do not need to be solutions to (3.1) of chapter I , we cannot use the the maximum principle in proving an inequality such as (3.27) of chapter I for the adjoint Green's. function. Also given (3.27) one appears to need Hölder continuity of the adjoint Green's function near ∂U in order to complete the proof of Lemma 3.4 in chapter I. Finally we point out (as mentioned in the remarks after Theorem 1.10 in chapter II and in the remark at the end of section 2) that Lemma 4.6, the elliptic version of Lemma 2.42, and [CF] imply a weak version of Theorem 2.5 in [FKP]. Still though one cannot use this theorem as in [FKP] to pass from the small norm R. Fefferman condition to the large norm R. Fefferman condition (Theorem 2.4 in [FKP]) because our basic estimates also depend on λ and not just on the ellipticity constants. Thus the extrapolation argument used in the proof of Theorem 1.7 is also needed in the elliptic case (Theorem 1.9).

5. References

[A] D Aronson, *Non negative solutions of linear parabolic equations*, Ann. Scuola Norm. Sup. Pisa **22** (1968), 607-694.

[CF] R. Coifman and C. Fefferman, *Weighted norm inequalities for maximal functions and singular integrals*, Studia Math. **51** (1974), 241-250.

[DJK] B. Dahlberg, D. Jerison, and C. Kenig, *Area integral estimates for elliptic differential operators with nonsmooth coefficients*, Ark. Mat. **22** (1984), 97-108.

[Fe] B. Fefferman, *A criterion for the absolute continuity of the harmonic measure associated with an elliptic operator*, J. A.M.S **2** (1989), 127-135.

[FS] E. Fabes and M. Safonov, *Behaviour near the boundary of positive solutions of second order parabolic equations*, J. Fourier Anal. Appl. **3** (1997), 871-882.

[FKP] B. Fefferman, C. Kenig, and J. Pipher, *The theory of weights and the Dirichlet problem for elliptic equations*, Ann. Math. **134** (1991), 65-124.

[N] K. Nyström, *The Dirichlet Problem for Second Order Parabolic Operators*, Indiana University Mathematics Journal **46** (1996), 183 - 245.

Steve Hofmann
Department of Mathematics
University of Missouri
Columbia, MO 65211-0001
hofmann@math.missouri.edu

John L. Lewis
Department of Mathematics
University of Kentucky
Lexington, KY 40506-0027
john@ms.uky.edu

Editorial Information

To be published in the *Memoirs*, a paper must be correct, new, nontrivial, and significant. Further, it must be well written and of interest to a substantial number of mathematicians. Piecemeal results, such as an inconclusive step toward an unproved major theorem or a minor variation on a known result, are in general not acceptable for publication. Papers appearing in *Memoirs* are generally longer than those appearing in *Transactions*, which shares the same editorial committee.

As of January 31, 2001, the backlog for this journal was approximately 7 volumes. This estimate is the result of dividing the number of manuscripts for this journal in the Providence office that have not yet gone to the printer on the above date by the average number of monographs per volume over the previous twelve months, reduced by the number of volumes published in four months (the time necessary for preparing a volume for the printer). (There are 6 volumes per year, each containing at least 4 numbers.)

A Consent to Publish and Copyright Agreement is required before a paper will be published in the *Memoirs*. After a paper is accepted for publication, the Providence office will send a Consent to Publish and Copyright Agreement to all authors of the paper. By submitting a paper to the *Memoirs*, authors certify that the results have not been submitted to nor are they under consideration for publication by another journal, conference proceedings, or similar publication.

Information for Authors

Memoirs are printed from camera copy fully prepared by the author. This means that the finished book will look exactly like the copy submitted.

The paper must contain a *descriptive title* and an *abstract* that summarizes the article in language suitable for workers in the general field (algebra, analysis, etc.). The *descriptive title* should be short, but informative; useless or vague phrases such as "some remarks about" or "concerning" should be avoided. The *abstract* should be at least one complete sentence, and at most 300 words. Included with the footnotes to the paper should be the 2000 *Mathematics Subject Classification* representing the primary and secondary subjects of the article. The classifications are accessible from `www.ams.org/msc/`. The list of classifications is also available in print starting with the 1999 annual index of *Mathematical Reviews*. The Mathematics Subject Classification footnote may be followed by a list of *key words and phrases* describing the subject matter of the article and taken from it. Journal abbreviations used in bibliographies are listed in the latest *Mathematical Reviews* annual index. The series abbreviations are also accessible from `www.ams.org/publications/`. To help in preparing and verifying references, the AMS offers MR Lookup, a Reference Tool for Linking, at `www.ams.org/mrlookup/`. When the manuscript is submitted, authors should supply the editor with electronic addresses if available. These will be printed after the postal address at the end of the article.

Electronically prepared manuscripts. The AMS encourages electronically prepared manuscripts, with a strong preference for $\mathcal{A}_{\mathcal{M}}\mathcal{S}$-LaTeX. To this end, the Society has prepared $\mathcal{A}_{\mathcal{M}}\mathcal{S}$-LaTeX author packages for each AMS publication. Author packages include instructions for preparing electronic manuscripts, the *AMS Author Handbook*, samples, and a style file that generates the particular design specifications of that publication series. Though $\mathcal{A}_{\mathcal{M}}\mathcal{S}$-LaTeX is the highly preferred format of TeX, author packages are also available in $\mathcal{A}_{\mathcal{M}}\mathcal{S}$-TeX.

Authors may retrieve an author package from e-MATH starting from www.ams.org/tex/ or via FTP to ftp.ams.org (login as anonymous, enter username as password, and type cd pub/author-info). The *AMS Author Handbook* and the *Instruction Manual* are available in PDF format following the author packages link from www.ams.org/tex/. The author package can be obtained free of charge by sending email to pub@ams.org (Internet) or from the Publication Division, American Mathematical Society, P.O. Box 6248, Providence, RI 02940-6248. When requesting an author package, please specify $\mathcal{AMS}$-L^{A}TEX or $\mathcal{AMS}$-TEX, Macintosh or IBM (3.5) format, and the publication in which your paper will appear. Please be sure to include your complete mailing address.

Sending electronic files. After acceptance, the source file(s) should be sent to the Providence office (this includes any TEX source file, any graphics files, and the DVI or PostScript file).

Before sending the source file, be sure you have proofread your paper carefully. The files you send must be the EXACT files used to generate the proof copy that was accepted for publication. For all publications, authors are required to send a printed copy of their paper, which exactly matches the copy approved for publication, along with any graphics that will appear in the paper.

TEX files may be submitted by email, FTP, or on diskette. The DVI file(s) and PostScript files should be submitted only by FTP or on diskette unless they are encoded properly to submit through email. (DVI files are binary and PostScript files tend to be very large.)

Electronically prepared manuscripts can be sent via email to pub-submit@ams.org (Internet). The subject line of the message should include the publication code to identify it as a Memoir. TEX source files, DVI files, and PostScript files can be transferred over the Internet by FTP to the Internet node e-math.ams.org (130.44.1.100).

Electronic graphics. Comprehensive instructions on preparing graphics are available at www.ams.org/jourhtml/graphics.html. A few of the major requirements are given here.

Submit files for graphics as EPS (Encapsulated PostScript) files. This includes graphics originated via a graphics application as well as scanned photographs or other computer-generated images. If this is not possible, TIFF files are acceptable as long as they can be opened in Adobe Photoshop or Illustrator. No matter what method was used to produce the graphic, it is necessary to provide a paper copy to the AMS.

Authors using graphics packages for the creation of electronic art should also avoid the use of any lines thinner than 0.5 points in width. Many graphics packages allow the user to specify a "hairline" for a very thin line. Hairlines often look acceptable when proofed on a typical laser printer. However, when produced on a high-resolution laser imagesetter, hairlines become nearly invisible and will be lost entirely in the final printing process.

Screens should be set to values between 15% and 85%. Screens which fall outside of this range are too light or too dark to print correctly. Variations of screens within a graphic should be no less than 10%.

Inquiries. Any inquiries concerning a paper that has been accepted for publication should be sent directly to the Electronic Prepress Department, American Mathematical Society, P. O. Box 6248, Providence, RI 02940-6248.

Selected Titles in This Series

(*Continued from the front of this publication*)

For a complete list of titles in this series, visit the
AMS Bookstore at **www.ams.org/bookstore/**.